# IN THE FOREST *of* NO JOY

ALSO BY J. P. DAUGHTON

*An Empire Divided: Religion, Republicanism, and the*
*Making of French Colonialism, 1880–1914*

# IN THE FOREST

## *of* NO JOY

THE CONGO-OCÉAN RAILROAD

AND THE

TRAGEDY OF FRENCH COLONIALISM

## J. P. DAUGHTON

**W. W. NORTON & COMPANY**

*Independent Publishers Since 1923*

Copyright © 2021 by J. P. Daughton

For information about permission to reproduce selections from this book, write to
Permissions, W. W. Norton & Company, Inc., 500 Fifth Avenue, New York, NY 10110

For information about special discounts for bulk purchases, please contact
W. W. Norton Special Sales at specialsales@wwnorton.com or 800-233-4830

Manufacturing by LSC Communications, Harrisonburg
Production manager: Beth Steidle

Library of Congress Cataloging-in-Publication Data

Names: Daughton, J. P. (James Patrick), author.
Title: In the forest of no joy : the Congo-Océan railroad and the tragedy
of French colonialism / J. P. Daughton.
Description: First edition. | New York : W. W. Norton & Company, [2021] |
Includes bibliographical references and index.
Identifiers: LCCN 2021000745 | ISBN 9780393541014 (hardcover) |
ISBN 9780393541021 (epub)
Subjects: LCSH: Chemin de fer Congo-océan—History. | Railroads—
Congo (Brazzaville)—History. | Railroad construction workers—Abuse of—
Congo (Brazzaville)—History.
Classification: LCC TF119.C75 D38 2021 | DDC 385.096724—dc23
LC record available at https://lccn.loc.gov/2021000745

W. W. Norton & Company, Inc., 500 Fifth Avenue, New York, N.Y. 10110
www.wwnorton.com

W. W. Norton & Company Ltd., 15 Carlisle Street, London W1D 3BS

1 2 3 4 5 6 7 8 9 0

*For Karyn, Nathaniel, and Henry*

O earth, cover not my blood,
and let my cry have no place.

—JOB 16:18

# CONTENTS

# LIST OF ILLUSTRATIONS

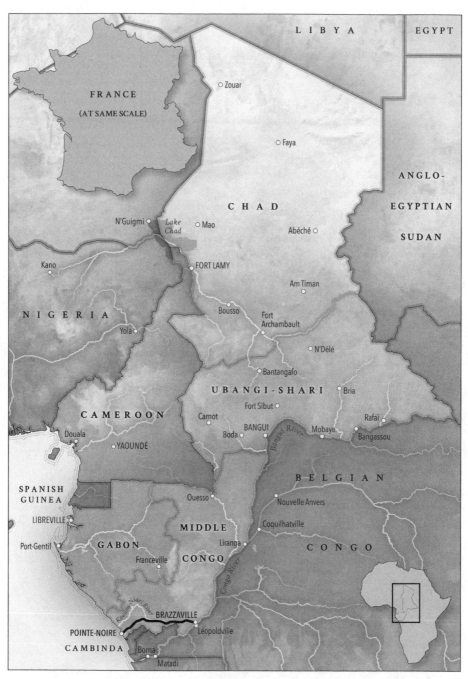

Map of French Equatorial Africa, c. 1920. The colony included the regions of Middle Congo, Gabon, Ubangi-Shari, and Chad (in French, Moyen-Congo, Gabon, Oubangui-Chari, and Tchad). The Congo-Océan railroad is at the extreme southern end of the colony.

# IN THE FOREST *of* NO JOY

INTRODUCTION

_____

# OF THOUSANDS GONE

Utopians are heedless of methods.

J. RODOLFO WILCOCK

ON NEW YEAR'S DAY 1925, Marcel Rouberol, the chief representative of the Société de construction des Batignolles, one of the largest
French engineering firms at the time, held a reception at his home
near Pointe-Noire for the men who worked under him. It was one
of many gatherings that company employees organized to fight the
boredom, homesickness, and sense of isolation that came with living
and working on an inclement expanse of sand thousands of miles
from home.

From 1921 to 1934, men from "the Batignolles" lived near the
coast in Middle Congo, often referred to by the French as simply "the Congo," a region in the southern part of French Equatorial Africa. They worked on building the Congo-Océan railroad, a
massive construction project that the colonial government undertook in the years just after the First World War. Long heralded by
Frenchmen as essential to the economic development of the region,
the railroad would connect the city of Brazzaville, the colony's larg-

Reception at the home of Marcel Rouberol, January 1, 1925. Rouberol is in the front row, third from left, and Mr. Martin, also in the front row, is fourth from left.

est settlement on the upper Congo River, to Pointe-Noire, on the Atlantic coast, where the French planned to build a deepwater port. Covering only some 512 kilometers, fewer than 320 miles, the railroad was not terribly long. But it crossed difficult terrain, especially the dreaded Mayombe, where the rails wound atop unstable, sandy soil, through a region of thick forests, mountains, and gorges.

A day like New Year's was worthy of a photograph—eighteen white men, all in pressed white linen suits, each with his white pith helmet in his hand. Corporations are built on hierarchies, so placement and positioning were essential in the picture, which was to be sent to company headquarters in Paris. The two men front and center are Rouberol and M. Martin, the lead engineer on the construction site. Around them are engineers, administrators, and overseers of the project. Two African auxiliaries, perhaps assistants or secretaries, lacking white linen and helmets, are wedged against the right edge of the image, one of them literally cut off by the frame, behind their white superiors.

The Batignolles men exuded confidence on that New Year's Day, none more so than Rouberol himself: his trimmed beard, neat tie, and impeccably white shoes complemented the self-assured smile on his face. Receptions like this one allowed these men and a handful of wives to come together and celebrate their accomplishments and discuss the work ahead. Despite their distance from France, executives of the Batignolles recreated some of the comforts of home. Boats from Europe brought in not only supplies to build the railroad but regular shipments of garlic, onions, and potatoes; refrigerated cheeses and charcuterie; jams and butter; rum, wine, and Moët champagne; Vichy and Perrier mineral water; coffee, Cointreau, and cigars. A New Year's party was just the place to enjoy all that civilization could offer.

Despite the inclement weather and prevalence of disease, the white men in the photograph, all well fed and standing tall, appeared to be paragons of camaraderie, cleanliness, and health. The photo was a testament to France's determination to triumph over the climate and landscape in a part of the world that Europeans considered insalubrious and uncivilized. The men of the Batignolles, as well as champions of the railroad in France, heralded the Congo-Océan as an engineering colossus achieved in the deadly wilds of the African continent. It would, as one newspaper put it, "save" Equatorial Africa, regularly called the "Cinderella" of the French Empire, and open the region's "nearly limitless" reservoir of riches.

While this portrait of colonial power is telling in many ways, there is another story that it does not tell—and that it in fact hides—in its straightforward simplicity. Just a short walk from the New Year's celebration was the construction site of the Congo-Océan where African men and women worked ten hours a day, six days a week, clearing millions of cubic meters of earth, building bridges

A "sick young woman and worker," c. June 1925.

and tunnels, and laying the ties and rails of the train line. For the thirteen years of construction, the workers wore minimal clothes and lived communally in huts so crowded and poorly ventilated that many chose to sleep outdoors. They faced overseers, both European and African, who often verbally tormented and beat them. They survived on a starvation diet and often went days without eating. Fresh water was in short supply, adding to the problem of dysentery that continually weakened the labor force, at certain points sickening or killing more than half the workers. Needless to say, French cuisine was never on the menu.

A second photograph, also taken in 1925 and not far from Rou-berol's home, brings the plight of these workers somewhat into focus. This image illustrates a very different side of the railroad project. The emaciated frames of these two unnamed people, identified only as a "sick young woman and worker," suggest they suffered from severe malnutrition and perhaps dysentery or beriberi, a thiamine deficiency common among the poorly fed recruits. The cachexia, or extreme wasting, of their bodies is reminiscent more of famine victims or prisoners of war than of what the French insisted they were—free, protected laborers in a republican empire.

Unlike the proud stances of the men of the Batignolles, these young people's bodies and faces betrayed uncertainty, humiliation, fear, and despair. It is impossible to know what exactly the "sick young woman and worker" thought about their predicament; the men of the Batignolles rarely documented the opinions of their workers in an effort to strip them of their voices and deny them their stories.

It is tempting to believe these two images came from different worlds, but they did not. Sent by steamer to France, the two pic-tures were placed in a book of photographs, not unlike an old family album, that the Batignolles kept in order to record the progress and history of the railway. The photograph of the two workers, with its unflinching documentary style, is reminiscent of many images used by humanitarian organizations to draw attention to atrocity and misery. The image, with the two young people framed by their background in a way that highlights their expressions and thin bod-ies, bears many hallmarks of what has been called the "aestheticiza-tion of suffering." But the purpose, or at least use, of this photograph was not to elicit sympathy or even pity. Instead, this collection of photos of workers, which were regularly left out of the company's publicly published material, filled out a private collection aimed at commemorating every aspect of the construction, from frustrations to daily habits to triumphs.

On the hunt between Pointe-Noire and Côte Mateva, August 1924.

Flipping through the Batignolles album reveals black and white photos of open plains, dense forests, rivers, and ravines; of laborers' huts, tunnels, bridges, a wharf, hand-pushed mine cars filled with dirt, and a newly built train station. There are photographs of the life of the construction for the Europeans, the white men and women carried by *tipoye* (sedan chair), or pushed in a wheeled *pousse-pousse* by African servants, or gathered for a concert in a high-ceilinged hall, or greeting the governor-general of the colony on a visit. It can make for jarring viewing. On one page, two white men and two white women out hunting strike poses worthy of a Tarzan film, their African guides standing to the side, visually fading into the background. And a few pages further on, a group of workers in the hot sun dig a massive trench without the help of machinery, pushing heavy wagonettes filled with dirt, under the watchful eye of European bosses. One, an image of leisure and adventure among

The construction site at Kilometer 53, March 1925.

the colonial elite; the other, a portrait of coerced labor. A caption of the first photograph identified the white men and women by name; in the second, the men remained nameless in a scene labeled plainly "The Work at Kilometer 53."

While such a juxtaposition seems striking now, at the time images of thin bodies, sick laborers, and men working in the equatorial heat were accepted parts of the daily life of the Congo-Océan. Many Europeans on the construction site thought little of workers' bodily conditions; indeed, most were convinced that Africans were better off for the work. In 1925 Gabrielle Vassal, the English wife of the French director of public health in the colony, wrote in her memoir that, while the workers' rations were "frugal, consisting chiefly of manioc, it is at least regular, and in this starving country keeps them as a whole cheerful and healthy." African malnutrition was a fact of life, she insisted, a by-product of inherent values. "The Congo population is always underfed," Vassal continued with an insight shared by many fellow Europeans, "and it is impossible to sound the depths of their laziness and want of thrift. They never even think of the very next day." The Congo-Océan—and all it represented about modern colonialism—would help rectify that.

Vassal's opinions, like the Batignolles album, are remarkable less for what they say about colonial hierarchies and racism than for their consistent denial of the full story of the construction of the Congo-Océan railroad. Histories of colonialism, especially in Africa before the Second World War, are rife with examples of European assertions of superiority in the face of allegedly savage subjects. The triumph of "civilization" over recalcitrant "primitives"—often bluntly portrayed in the European press as white men in pith helmets over naked, dark-skinned villagers—is a principal trope of modern imperialism. In this way, the Batignolles album was in keeping with European mentalities. What makes photos of champagne receptions juxtaposed with deprived workers particularly unsettling, however, is the casual coupling of the quotidian and the perverse. Recruitment to work on the Congo-Océan was widely recognized, by Europeans and Africans alike, as a possible death sentence. As André Gide, the future Nobel laureate in literature, put it after a visit to the French Congo in the mid-1920s, the Congo-Océan was "a frightful consumer of human lives."

Nearly entirely forgotten today outside central Africa, the Congo-Océan was one of the deadliest construction projects in history. The railroad likely caused between 15,000 and 23,000 African deaths, according to investigations conducted in the 1930s. Unofficial estimates were far higher, ranging from 30,000 to 60,000. In truth, the exact number of men, women, and children who died in the building of the Congo-Océan can never be known. French government statistics were not kept until a few years into the construc-

...et ça ne nous coûte que 2.000 « nègres » au kilomètre !!!

" . . . and that only costs us 2,000 'negroes' per kilometer!!!" *Le Journal du peuple,*
April 28, 1929.

tion, and even then they attempted to document only those workers who died on the worksite. The many thousands who died during recruitment itself, or who fled from the construction site never to be seen again, were not counted as casualties, even though many never made it home. Considering the serious administrative lapses evident during the construction from 1921 to 1934, even the statistics that do remain must be viewed with skepticism.

What is clear, however, is that the railroad was known, both in Equatorial Africa and in Europe, to be deadly. In France, its reputation gave rise to an oft-repeated assertion that the builders of the Congo-Océan laid as many corpses as railroad ties. In 1929 a biting cartoon from *Le Journal du peuple* showed a fat-cat capitalist, in white linen suit and with cigar in mouth, presenting the Congo-Océan, saying, "And that cost us only 2,000 'negroes' per kilometer." The steel rails stretch over the corpses of Africans.

Raw numbers of deaths fail to capture the full impact of the railroad on workers, their families, and their communities. One of the most troubling dimensions of the Congo-Océan was how prolonged and quotidian the mistreatment and misery became. The difficult working and living conditions, combined with thirteen years of sluggish progress, meant that the loss of life unfolded at a torturously slow pace. Hundreds died per month; thousands per year, year after year after year. Workers died at the hands of recruiters and overseers; they died in accidents and from wounds; they died of malnutrition, dysentery, and very possibly, from a disease unknown to doctors at the time but now called Acquired Immune Deficiency Syndrome, or AIDS.

In the history of construction projects, the mortality witnessed on the railroad in the French Congo has few peers. Even many premodern construction projects dependent upon enslaved labor rarely produced as many cadavers. In a little over a decade, more men and women died on the Congo-Océan than in eighty years building the Pyramids of Giza. In the modern era, railroad constructions around the world, including the building of thousands of miles of rails in the nineteenth-century American West, resulted in relatively few deaths compared to the Congo-Océan. Only the "French period" of building the Panama Canal in the 1880s witnessed comparable numbers of deaths—around 22,000—where workers fell to a repertoire of endemic diseases including yellow fever, smallpox, and typhoid fever.

In terms of twentieth-century comparisons, the Congolese railway can be likened to the most notorious cases of construction-related violence found in totalitarian and authoritarian regimes. While reliable figures are hard to come by, the sum of prison laborers who died in the early 1930s building the White Sea–Baltic Canal, or *Belomorkanal*, which is often cited as one of the most brutal chapters of Stalinist rule, was likely lower than the number lost in Equatorial Africa. Around twice as many Africans died building the Congo-Océan as British, Australian, and Dutch POWs succumbed working on the Burma-Siam Railway during the Second World War. Known as the "death railroad," the horrors of the Burma-Siam were made famous by the book and film *The Bridge on the River Kwai*, as well as by memoirs of survivors. Of course, exponentially more Asians than white prisoners died on that railroad, though their experiences have been far less often commemorated in books and film.

Key differences distinguish the case of the Congo-Océan from other episodes of brutality or slavery. Most important, African workers were not prisoners of Japanese militarism or Communist ideology. They were not slaves to pharaohs. Rather, they were subjects of a nation whose motto was "Liberty, Equality, Fraternity." All the same, many thousands were forced to work, usually at the end of a gun, stick, or whip. They were shackled together, made to trek for weeks on foot through swamps, deserts, and forests, and packed on crowded riverboats. Once on the site, they worked and lived in trying conditions where beatings were commonplace and payment, already meager, was often withheld for no justifiable reason.

Parallels to totalitarianism were not lost on men and women who had lived through the era of the Congo-Océan. René Maran, the French Guyanese writer who was the first person of color to win the prestigious Prix Goncourt, reflected after the Second World War that the "modern slavery" of the Congo-Océan had been "worthy of the worst Hitlerian practices." The immense colony, he noted, had been transformed into what would later be called "a concentration camp." While comparisons to the Shoah obscure the dynamics that allowed ostensibly liberal projects like the Congo-Océan to become sites of state-sponsored misery, Maran's provocative comparison should encourage a deeper look at colonial violence, its causes, and its impact on the men, women, and children who lived through it.

Put simply, the troubling details of the railroad compel explanation of *how* the Congo-Océan, with its widespread misery and catastrophic loss of life, happened. The railroad is not unknown to historians or curious writers; it has been the subject of scholarly papers and even glossy coffee table books. Surprisingly, though, very few have tried to explain the cruelty and brutality that defined so much of the project. None has attempted to explore how or why the French justified continuing construction in the face of such extraordinary human loss. It is a truism to say that modern European empires, especially in Africa, were racist, callous, and violent. So it is tempting to explain—and even dismiss—the building of the Congo-Océan as just another example of colonialism's penchant for inhumanity. But to understand life and death on the railroad, and in the colony that built it, we must dig deeper than racism, greed, and aggression, though all three clearly were present.

The Congo-Océan was not simply a case of brutal white men seeking a profit in a hidden corner of Africa. Instead, its construction marshaled the values, institutions, and know-how of a modern liberal state. While the era of the "technical expert" in Africa might still have been decades off, the 1920s were the starting point of a new appreciation for colonial policies that could improve diets, eradicate diseases, and build infrastructure. For the railroad's defenders, in addition to demonstrating the vitality of French imperial power, it was always a measure of French expertise, determination, and beneficence. Railroads had long been an enduring symbol of progress. And in the interwar years, the moving of mountains and building of bridges in the French Congo was deeply enmeshed in new ideas about alleviating suffering and developing societies not only in the name of empire but in the service of "humanity" more broadly. The railway would, as was often said, awaken the land from its "lethargic sleep" and bring its people's health, stability, and prosperity. The violence and mortality on the Congo-

Océan, then, cannot be understood without considering European claims to humanity.

European atrocities in Africa are often recounted as fables in which heartless villains are found, castigated, and punished by other, morally righteous Europeans. An influential model is Joseph Conrad's *Heart of Darkness*, a novella inspired by the author's time in Belgian King Leopold II's brutal Congo Free State (1885–1908). Conrad's narrator Marlow voyages upriver to retrieve Mr. Kurtz, a violent and delusional antihero who has strayed from Europe's supposed moral norms. The influence of Conrad's account can be seen in histories of Leopold II's oppression of the Congo. With its devastating system that forced villagers to harvest rubber or face rape, mutilation, or death, Leopold's Congo was ultimately denounced by Americans and Europeans who documented in government reports and the press what they had witnessed firsthand. Such a narrative of atrocity and reform, however, risks echoing the very rhetoric of Roger Casement, Edmund D. Morel, and other crusaders who initially exposed Leopold's brutality and believed it could be effectively reformed through political pressure.

The Congo–Océan railroad, by contrast, possessed few villains and heroes and witnessed far less inspiring reforms. A project spearheaded by administrators and company men, the Congo–Océan revealed the devastating potential of official reports, regulations, and spreadsheets to obscure and justify oppression, mistreatment, and even murder. While individuals and ideologies certainly colored the drama, the colonial state and its bureaucracy shaped the terms of debate about the railroad in important ways. Institutions, be they corporations, universities, or governments, try to craft their own narratives. They are equipped with protocols and hierarchies that ostensibly allow them to investigate and monitor their successes and failures, as well as the behavior of their personnel. They often pride themselves on adhering to concepts like "ethics" and "justice";

guarding supposed values and reputation is the paramount concern of most institutions. Self-preservation is too often a greater priority than admitting fault and self-improvement.

In the case of the administration of French Equatorial Africa, officials went to extraordinary lengths, on the one hand, to investigate and document and, on the other, to deflect and deny accusations of mismanagement and brutality. Their actions were shaped by practical concerns and personal convictions, as well as external pressures caused by monetary and material shortages, metropolitan politics, outspoken colonists, African workers, and the international press. The Congo-Océan, then, offers an instructive case of how a colonial state, shaped by the actions of its officers and the lenses of its bureaucracy, could become so misguided, indifferent, and even deluded, in pursuit of a project of paramount importance. Unbridled racism and greed, rather than causing colonial violence, informed the relationships, perspectives, and bureaucratic practices that ultimately made colonial violence appear justifiable and even necessary. There is something comforting in believing that hateful madmen made empires violent. In fact, negligence, denial, and assertions of humanity in pursuit of "progress" often proved far more cruel.

―――――

IT MAY BE TEMPTING, when reading the stories of the men and women "recruited" to build the Congo-Océan, to connect their experiences with other, perhaps better-known histories. For example, details of how railroad workers were often hunted, tied together in coffles, and beaten for not working hard enough encourage analogies with the long history of chattel slavery in the New World. French businesses, especially in the Caribbean, relied heavily on and profited handsomely from enslaved labor, from the seventeenth century until slavery's final abolition in France in 1848. Critics of the Congo-Océan certainly highlighted both the tragic and the ironic continuities between the transatlantic slave trade and the building of the railroad. Rife with brutality, coercion, and death, the Congo-Océan passed through lands haunted by memories of the transatlantic trade. Pointe-Noire, the western terminus of the Congo-Océan, sits less than a dozen miles south of the Loango harbor where, for centuries, millions of captured men and women were sold to European slavers. The fact that the railroad, like many projects in the age of "new" imperialism, promised to bring "progress" and "civilization" on the backs of free labor only made the similarities to slavery all the more jarring.

Other comparisons prove equally enticing. The stripping of dignity, the lack of food and medical attention, and the housing of workers in prison camps are at times reminiscent, as Maran pointed out, of the gross injustices of twentieth-century totalitarian regimes. The violence of the Congo-Océan was definitely murderous, but in ways that stray from totalitarian state violence. It was not part of a grand exterminationist scheme designed to destroy a race or a people. Unspeakable suffering was not wrought by secret police or ideologues; rather, it stemmed from weakness, failure, and mismanagement. The railroad's brutality was, in turn, petty, unthinking, and unconcerned, justified by racist beliefs that conveniently displaced moral responsibility.

Analogies are far more effective rhetorically than historically. For all the apparent parallels, the legal bases, economic motivations, and human experiences of chattel slavery differed starkly from the labor system found on the Congo-Océan. The same could be said of the use of prison labor in 1930s Europe. More concerning still, when these sorts of analogies are deployed as more than provocative metaphors, they can prompt comparisons of human suffering that do little to enlighten the specific contexts and experiences of the men and women who endured the construction of the Congo-Océan. Exploring continuities and parallels is certainly essential to understanding the past; but there is also the danger of overlooking pervasive and instructive phenomena—in this case, the violence of empire—for not being "as bad" or "as significant" as other cases. Seeing colonial violence primarily as a legacy of slavery or as a prelude to the "crime of crimes" of genocide can diminish its historical relevance. The aim of studying cases like the Congo-Océan is not to establish rankings; rather, it is to better understand how modern, ostensibly liberal European regimes established and legitimized their rule over vast overseas communities, and how those communities experienced it.

The plight of the men and women recruited to build the Congo-Océan was not hidden in the deserts of Chad or the dark forests of the Mayombe. Their experiences were well documented by an array of witnesses at the time. Conditions were repeatedly debated in the halls of government. The construction was described in books and illustrated in newspapers, often in accounts penned by some of the era's leading writers. Thousands of pages of official reports recorded descriptions of conditions, incidents of brutality, and the raw data of the number of people who perished from disease and neglect. Libraries and archives hold an abundance of materials, from narratives by firsthand witnesses to testimony given by African workers themselves, that make vividly clear the hardships endured. But as the Batignolles photo album makes clear, there were competing narratives. Defenders of the railroad, including the colonial administration in Equatorial Africa, insisted that the privation and deaths on the railroad were grossly exaggerated or entirely justifiable.

The rich material provides panoramic detail about what happened, from the deeds of pro-colonial Frenchmen to the experiences of people, like the sick young woman and worker, on the Congo-Océan. While the former are easy to find if not always to understand, the "voices" of the workers are notoriously more difficult to hear, perhaps especially considering that their experiences included fundamental passages in life, like loss of family, physical pain, illness, and death. The hope, agony, and fatigue of African men and women who often did not write, and who rarely spoke a language that any Europeans on the construction site could understand, fall silent in the cacophonous accounts of the construction. But their actions are not absent; nor are their experiences lost.

Government documents include interviews, depositions, and testimony of African workers that provide telling insights. African clerks, though few in number, editorialize in their reports, offering unexpected reactions to events. Former workers left oral accounts many years after the fact. An array of detailed secondary accounts of working conditions, written by European journalists, travelers, and officials, including direct encounters with recruits, laborers, and deserters, are equally rich sources. There are films, photographs, and drawings from the period. These sources can be combined with phys-iological information, based on food rations and data about workers' weight, health, and living conditions, as well as geographical and climatic studies of the places where the drama played out. While the thoughts of individuals and specific moments are often lost, the conditions in which bodies and minds survived or perished are not.

Like any information historians use to recreate the past, the Congo-Océan's sources are imperfect and incomplete. They were produced overwhelmingly by white men and women for the use of Europeans. Some representations were deeply shaded by politi-cal perspective, economic interest, and personal history; they could range from the purposefully misleading to the cynically false. Some sources provide only the broadest of generalizations—for example, food on the railroad was good and plentiful, or it was rancid and scarce—while elsewhere in the archives, the lives of specific indi-viduals come into extremely tight focus. Much can be learned, for example, of certain workers who stood accused or who were the victims of brutality; officials recorded their names, their home vil-lages, their parents' names, their reputations, or their whereabouts on a certain afternoon when an incident occurred. They appear for a fleeting moment, never to be mentioned again.

By contextualizing, comparing, interrogating, and questioning these sources; by balancing individuals' opinions and experiences with general impressions; by turning archives back on themselves, using them to reveal their own strategies or means of knowing—by employing an array of techniques, it becomes possible to document uprooted lives, to perceive the concerns of men and women who had been taken far from home, and to begin to comprehend the momentousness of the choices they were often forced to make. Pain and suffering are notoriously difficult to measure and describe. Bodily trauma renders language imprecise and unstable; it tends to elicit discussion of comparative miseries that do little to capture and much to undermine comprehension of the trauma at hand. And yet it remains essential to use violence and suffering as windows onto the Congo-Océan. Both contributed to arguments about the humanity or inhumanity of the project, shaped policies taken by the colonial state, and influenced the behaviors of workers on the construction site.

If these sources need scrutinizing like any other, drawing on them to explore African experiences, attitudes, and actions is also essential. Attention to what historians call the agency of these men and women can help expose, if not rectify, past injustices that continue to color memories of colonialism as well as policies in the postcolonial world. Many accounts of the Congo-Océan recorded during and immediately after its construction purposefully minimized or effaced Africans' role. As in the Batignolles album, Africans were often reduced to "types" or "specimens" of nameless men and women. When African workers provided eyewitness testimony, their accounts were often dismissed as unreliable and prone to falsehood. The same skepticism, shaped in no small part by racism, also spread to international organizations like the League of Nations.

By marginalizing the workers on the Congo-Océan—and indeed on all colonial projects in which local communities contributed with sweat and blood—historians risk perpetuating the same misleading narratives that helped justify the projects in the first place. The ideology of modern imperialism, with its calls for modernized infrastructures, capitalist development, and liberal ideals, measured success in terms of profits made and challenges overcome. Such a history was foundational of national and racial identity in nineteenth- and twentieth-century France and Europe more broadly. But this history also erases the consequences for the men, women, and children who lived under French rule. To draw a parallel with Ned Blackhawk's assessment of American history, marginalizing the experiences of workers on the Congo-Océan means allowing European history to be "a realm of achievement rather than one of indigenous trauma."

Many history books, as well as documentary and fictional films, have endeavored to explain the root causes of abuses and atrocities committed by totalitarian and authoritarian regimes around the world. Historians of modern Europe have been committed to understanding how civilian neighbors and ordinary men in uniform were motivated to murder in the service of ideologically driven and genocidal regimes. By contrast, instances of colonial violence, often perpetrated not by uncontrolled mobs or soldiers in jackboots but by men working for international engineering companies or serving official roles in ostensibly liberal governments, remain not only neglected but also forgotten. Outside the Republic of Congo, the deadly history of the Congo-Océan is as unknown to the general public today as the railroad itself.

Such willful forgetfulness can no doubt be attributed in part to the fact that nations don't like to remember their more unsavory periods of history. The Congo-Océan, however, requires reassessment of what has been denied or forgotten and to what end. Supporters of the railway—the men at New Year's parties and ladies out hunting, as well as businessmen and politicians in Paris—went to great lengths to deny or justify the deaths of thousands of workers. Their methods were deeply infused with racism and concerns about profits. But they also had to devise other ways to hide or mitigate the realities of the railroad. The means that helped Europeans morally absolve themselves of guilt for inflicting a multitude of privations on their colonial subjects are not simple relics of the past; they also laid the foundation for the forgetfulness of such histories today.

The story that follows includes graphic and detailed accounts of human cruelty. Unlike many sites of great violence, where perpetrators were only too happy to cover up their deeds, the Congo-Océan was documented with all the procedures of a modern bureaucracy. As a result, the archives left to historians contain strikingly cold and concise records of abuse, ranging from beatings and torture to starvation and death by disease. These reports were produced by Europeans, but they often incorporate and even quote the words of those who were forced to work. Most of the details of these accounts have been left unexplored and uninterrogated by historians. Unlike other histories of atrocities, in which bearing witness is an integral part of recording the past, colonial violence is mentioned in passing if at all.

Recounting the experiences of men and women—whose names were often recorded—contributes to a necessary assessment of what the violence of empire can and did look like. It bears witness to a chapter of global history that many, especially in Europe and the United States, have endeavored to deny and forget. Historians have produced many moving and troubling accounts of the details of European lives gripped by poverty, of European refugees, of European prisoners who endured unimaginable torments at the hands of modern governments. By contrast, the details of imperialism, a historical upheaval that profoundly forged the modern world, are often left to hover in the dim margins of history.

Describing the details of men being beaten or tortured might be construed as sensationalist or as reveling in a kind of "pornography of pain." Many scholars fear that Africans are too commonly portrayed as passive victims, and that the history of Africa focuses too much on what was done to it, rather than on the accomplishments of its people and cultures. There is great merit to these concerns. But recounting the story of the men and women who built the Congo-Océan is not an exercise in victimization. It writes their extraordinary work, perseverance, and trauma into the history of modern Europe and imperialism, as well as the global history of humanity and humanitarianism. This African railroad, with its miseries and its feats, is a story relevant far beyond the three hundred miles its covers.

Indeed, telling the story of the Congo-Océan requires asking why it hasn't been told before, at least in a form that accounts for the myriad trials faced and the momentous choices made by the various actors in the drama. And relatedly, why are stories of other kinds of violence so much more familiar? What do histories of colonial encounters offer to the great repertoire of human experiences, vices, and virtues that have made them less worthy of telling? Why have so few historians tried to bear witness—an act so often deemed essential to modern life—to these pasts?

Here the Batignolles album, with its photographs that eerily defy time, encourages a kind of reckoning. The men sweating in the afternoon sun, the sick woman and the worker—their faces are buried now, flattened in dark albums, hidden in file boxes in guarded archives in secondary European cities. But the faces of the men and women and children, when called on, still gaze out. They beseech attention. They *oblige* history.

## CHAPTER 1

—————

# REMAKING THE CONGO

They used to say:
"They come from the bottom of waters, from the shadow-
lands where the dead live . . ." or again: "They came in great
whales blowing smoke, they came up from under the ocean
where the spirits live."

EMMANUEL DONGALA,
LE FEU DES ORIGINES (1987)

THE VAST BASIN AROUND the Congo River was, for many Europeans, an abstraction, a state of mind, a myth, a collection of
fantasies, an idea, a blank space to be filled in. Since the early sixteenth century, Europeans had thought of equatorial Africa as a
place of mysterious peoples, despotic chiefdoms, cannibalism, and
the slave trade. It was a forbidding territory of impenetrable forests
that swallowed explorers alive or laid them low with deadly diseases. At the nineteenth century's close, Joseph Conrad solidified
and modernized this image in vivid prose of folly, brutality, and
gloom. "Going up that river," his narrator in *Heart of Darkness* says
of the Congo, "was like travelling back to the earliest beginnings

of the world, when vegetation rioted on the earth and the big trees were kings."

Of course, late nineteenth-century Europeans were ignorant—in most cases willfully so—of the societies encountered there. The vast region that would become French Equatorial Africa had been home to kingdoms, sultanates, and chiefdoms for more than two millennia. The Kingdom of the Kongo, established at the end of the fourteenth century, covered territory south of the river along the coast in modern-day Angola, as well as inland in what is now the Democratic Republic of Congo (Kinshasa) and the Republic of Congo (Brazzaville). To the north, three smaller kingdoms—of Loango, the largest, as well as Kakongo and Ngoyo—prospered from the sixteenth to the nineteenth centuries. Polities rich in agricultural and commercial goods, including ivory and copper, as well as exquisite artifacts like pottery, cloth, and carved pieces, were linked by trade networks that ran along the coast and inland up the Congo River. Early European visitors found large towns with monarchical courts, massive caravans transporting goods, and vibrant markets, like the one at Pool Malebo, frequented by dozens of boats.

To the north of what became French Equatorial Africa were the even older kingdoms, often called sultanates, of Kanem-Borno, Wadai, and the smaller, weaker Bagirmi. Kanem-Borno (also called Bornu) dated from the eighth century, Wadai and Bagirmi from the late fifteenth. These kingdoms were important political and military powers along trans-Saharan trade routes that stretched from West Africa to the Sudan and almost to the Red Sea, and north to the shores of the Mediterranean. As on the Atlantic Coast, enslaved people were the preferred item of sale, with millions of men, women, and children forced to migrate along the trade routes. By the nineteenth century, the region had established diplomatic ties, as well as a military alliance, with the Ottoman Empire.

In 1910 the individual possessions of Gabon, Middle Congo

(Moyen-Congo), Ubangi-Shari (Oubangui-Chari), and Chad (Tchad) were unified as French Equatorial Africa, a massive territory that encompassed the four present-day nations of Gabon, Republic of Congo, Central African Republic, and Chad. The regions shared little in common ethnically or environmentally. The rain forest of Gabon and Middle Congo gave way to savannahs in Ubangi-Shari and drier pastures in the north, where Chad's landscape blended with the Sahara. These regions had their own corresponding agricultural traditions, ranging from hunting and gathering in the rain forest to cattle herding in the north. In terms of climate, geography, flora, and fauna, the north and south of what became French Equatorial Africa could not have been more different. These varying geographies possessed a variety of ethnolinguistic groups, including many Bantu speakers in Middle Congo, Gbaya (Baya) and Banda in Ubangi-Shari, and Sara, Baguirmi, and Arabs in Chad; and there were nomadic groups that defied colonial borders entirely. While trade between subgroups was common, many possessed unique linguistic, political, religious, and cultural traditions.

Far from being cut off from the outside world, the kingdoms and small chiefdoms of central Africa traded goods and established diplomatic ties with Europeans and the Ottomans and became major sources of both the transatlantic and Muslim slave trades. This participation ultimately greatly undermined political and economic stability across the region. The polities of central Africa had certainly not been idyllic before the arrival of Europeans. Domestic slavery and violent conflicts between rival groups were not uncommon. But before the intensification of the international slave trade, domestic slavery was largely limited to people taken in battle, and wars were more often about relatively limited competition than about territorial expansion. The European thirst for human chattel between 1650 and 1850 pushed African slavers farther into the interior, causing dislocation as communities either fled or fought the invaders. Regional factions violently turned on one another in the search for men and women to sell, tearing apart kingdoms and communities alike.

Equatorial Africa played an important role in the making of the modern world, with commercial, political, and human connections to Europe, the New World, and Muslim powers to the north and east. On the eve of French colonization, this history had caused significant migrations and reshuffling of peoples. The Kongo and Loango kingdoms had lost their influence; individual chiefs vied for power among followers who'd grown wary of centuries of raids and conflict. The central Sudanic kingdoms of the north were unable to stop French conquest when it came after 1900. Europeans preferred to imagine the African societies they encountered as being as wild and as old as the primeval forest rather than a by-product of their own slave-trading past. In the age of high imperialism, European colonizers found it difficult not to view the people they claimed as subjects through the distorted lenses of race and civilization. For them, the indigenous communities were inherently gripped by poverty, despotism, and despair, victims of their own alleged immorality, ignorance, and depravity.

Distortions and prejudices thus informed European ideas about what Africa needed. Africans were an obstacle to be overcome. Their societies were as wild as the landscape. The forests, plains, mountains, and serpentine rivers; the sun, the heat, the humidity, and the storms—all stood in opposition to Europe. Here both man and nature challenged self-assured Europeans, even called out to be conquered, tamed, harnessed, civilized, developed. If the Congo were to be exploited and made habitable—that is, by and for white men and women—it would have to be remade. The early efforts to remake the regions that would become French Equatorial Africa provide the context in which the building of the railroad took shape. Early French approaches also foreshadowed many of the themes— political ambivalence, poor administrative organization, violence, the rhetoric of humanity, and an apparent distaste for advance planning—that would plague nearly every stage of the construction.

———

IF ONE FIGURE FORMED a bridge between the initial French exploration of equatorial Africa and the eventual decision to build the Congo-Océan, it was the Italian-born, naturalized Frenchman Pierre Savorgnan de Brazza. The son of an Italian aristocrat with solid political connections, Brazza joined the French Navy as a young man and soon developed a taste for exploration. In the mid-1870s, at barely twenty-two years of age, Brazza undertook several expeditions with the financial backing of his family and a number of prominent French interests. He first ventured into the interior of Gabon, a region largely unexplored by Europeans, heading up the Ogooué River, which local sources claimed cut deep into the continent. Three years later Brazza had discovered that the Ogooué did not lead to the heart of Africa, but he had a good sense of how he could cross from Gabon to the northern bank of the Congo River.

As Britain, Belgium, Portugal, and Germany were all backing exploratory projects in parts of central Africa, Brazza's expeditions transformed him into a national hero and immediately lent his actions political significance. In 1880 he negotiated a treaty with Iloo, the *makoko,* or leader, of the Téké people and an influential regional ruler, to stake a claim to a French protectorate over a vast territory north of the Congo River. In a sign of Brazza's distance and rogue diplomacy, it was months before anyone in Paris even knew of the treaty and two years before the National Assembly ratified it. Even then questions remained, not least of which was whether Iloo and Brazza shared a mutual understanding of the terms of the papers signed. The agreement was deemed good enough in an age of colonial expansion in Africa. Politicians in distant Paris claimed the territory north of the river as *le Congo français*—the French Congo.

While Brazza shaped France's presence in equatorial Africa, his *image* came to define how many French people viewed their new colonial venture. He was an almost instant hero, not simply for negotiating the acquisition of a massive new territory but for having an unexpected persona as well. The reigning colonial adventurer of central Africa in the 1870s was Henry Morton Stanley, the Welsh-born explorer, journalist, and notorious self-promoter. A "superhuman hero," as one French writer called him, Stanley first gained international fame by finding the lost missionary David Livingstone; his line "Dr. Livingstone, I presume," allegedly spoken in a small village on the coast of Lake Tanganyika after eight months of searching, was reproduced in papers around the world. By the time of Brazza's exploration, Stanley was busy helping Belgian king Leopold II make the Congo Free State—the region largely south and east of the Congo River—into a profitable enterprise. Stanley exuded the masculinity of his era, never shy to promote his own bravery, brashness, and swagger. By contrast, Brazza bucked the rough-and-tumble image of a conqueror, portraying himself instead as a wanderer, a prophet, and a mystic, or as Stanley called him with as much envy as sarcasm, "an apostle of liberty, the new apostle of Africa."

Images of Brazza—the most famous of which was made by the legendary portrait photographer Nadar—showed him barefoot, in torn clothing, his head wrapped in a tunic, his eyes fixed thoughtfully on a distant horizon. He carried not a gun but a staff—the staff of a shepherd that he used at once to expand France's reach in the world and to guide its new subjects on the right path. As the historian Edward Berenson has pointed out, in the 1870s, in the wake of France's humiliating defeat in the Franco-Prussian War, the French press was eager to find new emblems of the nation's ideals. Brazza was presented as a peaceful conqueror who brought freedom and civilization to the people he "liberated." He appealed to a wide spectrum of Frenchmen, from the ardently secular who admired his republi-

Pierre Savorgnan de Brazza (1905), portrait by Nadar.

can ideals to the devoutly religious—both Catholic and Protestant—
who wanted France to evangelize abroad. Brazza captured perfectly
what many pro-colonials liked to imagine was France's benevolent
determination.

Brazza's appeal stemmed from his stated commitment both to his adopted nation and to humanity. He was not motivated by humanitarianism per se—that is, he did not expect his countrymen to invest in colonization because of an abstract moral compulsion to improve or save African lives. He was, instead, very much a creature of his era, a colonial advocate who encouraged France to pursue its "civilizing mission" in regions of the world that Europeans deemed regressive. "Civilization" promised to bring all the trappings of science, technology, and enlightenment to the less fortunate; the promotion of a particularly French vision of civilization also demonstrated the nation's vitality, global standing, and liberal magnanimity. In 1880, for example, he bought slaves in Gabon their liberty, declaring before the *tricolore* of the French Third Republic, "All those who touch our flag are free."

Brazza made *humanité* a pillar of his rhetoric and style. He often used the word in reference to his own thinking; it even crept into his letters to his wife. Such language was not entirely new. Allusions to humanity in colonization harkened back to the language used by mid-nineteenth-century romantic socialists to justify French rule in Algeria, which they imagined as freer and more "civilized" than the slave economies of the "old" Caribbean colonies. But Brazza's ideas of humanity departed from ideology: it was in his personal behavior as an explorer that his self-styled humanity took shape. While on expedition, he strove to avoid violent conflict with even the most warlike populations he encountered. His strategy, necessary considering his long voyages with few soldiers, was to convince local chiefs that their interests coincided with French ones and to make treaties accordingly.

In an 1886 speech given in France, Brazza boasted of winning over all the tribes along the Ogooué River in Gabon, even the allegedly dreaded Pahouins. "These cannibal tribes," he said, with "their savagery" and "their instinct for pillaging," were made to see the benefits of working with the French. They began to trade, provided porters and guides, and even convinced other recalcitrant groups of the benefits of French influence. It was "a considerable guarantee from the point of view of tranquility," Brazza said. "It was maybe the only way of maintaining complete security in a country that is absolutely—I was going to say happily—outside the reach of gunboats." Speeches such as this one, at times delivered to crowds of thousands, solidified his reputation. As one commentator in 1887 put it, equatorial Africa was "developed thanks to [Brazza's] tenacity, his patience, his indomitable perseverance," by "exclusively peaceful means."

By the early 1880s, Brazza had already started to promote the construction of a railway connecting Pool Malebo, the site of Brazzaville and Léopoldville, the largest European settlements in the region, to the Atlantic Coast. He was not alone. "Without a railroad," Stanley was often quoted as saying, "the Congo isn't worth a penny." It wasn't difficult to understand why both men were so preoccupied with rails. Pool Malebo (which Europeans dubbed Stanley Pool) is a calm, wide, lakelike expanse of water that marks the last navigable stretch of the lower Congo River for boats arriving from the continent's interior. Just below Pool Malebo, toward the ocean, lay a series of impassable rapids and falls. Goods brought downriver had to be offloaded at the pool and taken overland, on porters' backs, across difficult and deadly terrain to the coast, where they could be shipped on to Europe or elsewhere. Brazza believed a railroad would open the many societies of equatorial Africa to global markets.

Brazza's motivation for supporting the construction of a French railroad in equatorial Africa was never entirely about making money; it rather included the development of the region and the salvation of the allegedly suffering peoples of the Congo and beyond. Brazza wanted to make the Congo both a productive and a profitable possession, as well as a central component of a larger French empire. As early as 1883, he assured his compatriots that the "enormous riches" of the Congo awaited only a railroad to deliver them. A rail line would guarantee French businessmen access to "the vastest ensemble of navigable rivers" in all of Africa. The Congo and Ubangi basins, at more than 4 million square kilometers (an area larger than India), promised to yield resources big and small, from rubber and ivory to manioc and pistachios. Echoing British calls to link the Cape to Cairo by rail, Brazza was already imagining a railway to be part of a much larger system of French "routes of penetration" into and across the continent, stretching from the equatorial western coast all the way to Egypt, with Brazzaville a key hub.

Brazza's vision to remake equatorial Africa showed extraordinary audacity. He touted a united French Africa, stretching from the mouth of the Congo River all the way to French colonies in West and North Africa. And to this end, he promoted further expeditions. In 1887 the French staked a claim to Ubangi-Shari in the northeast. Expansion to the northwest, by contrast, was hampered by British and German claims in Nigeria and Cameroon respectively, as well as by a regional leader and slave trader named Rabah. While the French occupied territory on the eastern shore of Lake Chad, the boundaries would not be solidified until 1900. Having negotiated territory with alliances and treaties, Brazza was rewarded by his adopted nation in 1886 with the post of *commissaire-général,* the ranking administrator in the new possession.

Brazza tried to infuse his administration with the same humane aspirations that had guided his earlier explorations. In some ways he succeeded: territories in Ubangi and Chad were secured through diplomacy, negotiation, and relatively little bloodshed. Like many a great visionary, however, he was not the most effective of administrators. The fact that French rule in the region remained poorly defined and funded did not help. Brazza's administration floated in a fragmented and disorganized space within the bureaucratic hierarchy of metropolitan France. His office was politically, financially, and administratively directed not by a single ministry but was linked to at least three: foreign affairs, public instruction, and navy. Such broad ministerial attention might have meant overlapping support, but it didn't. The tiny budget the French government earmarked for the Congo mainly supported expeditions in Ubangi-Shari and Chad. Across the possession, a handful of stations and posts were opened, but they remained isolated and understaffed.

The reigning ideology of the era, exemplified by France's great colonial champion, Paul Leroy-Beaulieu, was that colonial development could and should, through production and taxation, essentially pay for itself. Politicians in Paris hoped new business ventures would be launched in equatorial Africa by merchants reassured by the stability promised by an official French presence. But the merchants did not materialize. Brazza tried to corral interest more directly, but he had difficulty finding serious investment. His hope of transforming the Congo into a profitable colony remained out of reach, in a part of the world lacking infrastructure, capital, and Europeans. Attracting more speculators, then, became increasingly connected to building a railway that would link the region's natural wealth to markets abroad.

Building a railway sounded good in theory but proved elusive in practice. The 1880s and '90s saw a number of efforts to establish a route for a railroad from Brazzaville to the coast. As early as 1882, Brazza had suggested a combined rail and river route across the plains due west of Stanley Pool to the Kouilou-Niari River system and down to the coast. In 1887 a midlevel officer named Pleigneur drowned while working on the Niari River, a sign of the perils of early exploration. Léon Jacob—a true nineteenth-century colonial type, a low-level administrator who fancied himself an adventurer, a cartographer, and perhaps most important, an engineer—took over. He spent months traversing more than four thousand kilometers of difficult terrain through equatorial jungle, mapping possible routes.

Jacob's attitudes reflected much of the hubris that would inform the building of the railroad. He insisted that his mapping—with compass, pedometer, and astronomical calculations—accounted for "all the twists and turns of the route" with "serious guarantees of exactitude." He admitted that the exactitude of the names of villages on his maps could not be verified. But finding blame was never hard: "the natives, very suspicious, often deliberately provided false leads."

Despite Jacob's self-assurance, none of the routes he recommended were without significant problems. One of the most plausible plans included rendering the Kouilou-Niari River navigable by building a series of dikes, thereby flooding the rapids that caused boats to run aground. A railroad would then be built to cover the rest of the distance—about 125 kilometers—to Brazzaville. The plan had some benefits: it was cheap, and it wisely avoided passing through the most mountainous area between Brazzaville and the coast. Unfortunately, though, the Kouilou-Niari presented challenges beyond the rapids, including sandbars near its mouth that were largely impassible to oceangoing vessels. Sealing its fate was the fact that the government was already investing in rail projects in West Africa and Indochina, leaving the till empty for the Congo. Other missions followed suit, but none could solve the intractable challenges of topography and lack of funds.

Meanwhile, as the French mapped the bush and debated potential budgets, across the river in Leopold's Congo, the Belgians were already building. In 1898 construction was completed on a 365-kilometer railroad from Léopoldville to Matadi, a port with open access to the sea. Though the railroad was owned by France's colonial rivals, it offered French producers an outlet to global markets. Political apathy in France about funding its own project at the cost of tens of millions of francs encouraged most colonial interests to content themselves with using the Belgian rails. The same year witnessed Brazza's departure from the Congo administration. The word from Paris had remained largely unchanged from the early days of his expeditions: as one minister had told him, "our interests are relatively weak" in the Congo. His aim of transforming the Congo into a profitable and peaceful colony remained woefully unfulfilled. And the advent of a new approach to colonizing would set his hopes back even further.

———

BRAZZA'S PLAN TO SERVE humanity through colonization was never implemented. His tenure had failed to build infrastructure, from telegraph lines to roads, or even to produce many studies of the population or the topography. Upon returning to France, he was accused of budgetary negligence, leaving the Congo administration in need of a government bailout, speeding his own dismissal. But beyond overspending the possession's means, the explorer and *commissaire général* never formulated a convincing economic plan. He promised that extraordinary wealth in human and material resources awaited investors, yet he was deeply skeptical of free trade, fearing that competition among merchant houses would drive down prices of goods and labor, discouraging investment and harming Africans. Skepticism about profitability continued to reign until the late 1890s, when a rubber boom led to hitherto-unknown levels of speculation and exploitation.

Starting in 1898, the continued lack of political interest in equatorial Africa encouraged lawmakers in Paris to adopt a means of development practiced in other European empires: the concessionary system. Under the system, the French government awarded land concessions to companies hoping to exploit natural resources. At the time, the main draw was wild rubber, with ivory, lumber, and palm oil less attractive but still profitable commodities. The government ceded tracts of land to companies to exploit. Following international agreements established at the 1884–85 Berlin Congress, concessions could not exclude other merchants from setting up shop on their territory. But the conditions of the concession granted the holding company exclusive rights to do business with local African producers, quashing competition. As a result, the concessionary system essentially created monopolies whereby a company's merchants purchased goods from local producers for export to Europe.

For a country unwilling to subsidize colonial economies, the concessionary system promised great benefits—most notably, it was colonizing on the cheap. The French state expected companies to develop their holdings with roads and lines of communication, establish a physical presence in their regions, and take up some of the work of the nation's supposed "civilizing mission" by encouraging work, education, and hygiene. Economically, the system had produced unexpected profits just to the south, in Leopold II's Congo Free State, where more than sixty-five concessions had a value of nearly 250 million francs. In the Congo Free State, concessions had seen production, profits, and stock values increase exponentially in less than five years. To politicians in Paris, concessions seemed ideal, the ultimate way to make colonialism pay for itself and then some.

By the end of the century, few people had expressed concerns about extending concessions in Gabon, Middle Congo, and Ubangi-Shari. Some forty companies came to occupy the majority of the area of the possession. The sheer size of some of the concessions was astounding, with the largest covering areas akin to midsize European countries. The Société du Haut-Oguoué, in eastern Gabon, one of the first concessions, occupied an area about the size of Iceland. The main concession in the Upper Ubangi covered a region larger than contemporary Greece. Even smaller concessions were still the size of American states like Delaware, Connecticut, and New Jersey. Companies acquired these territories with little or no knowledge of local communities and with no infrastructure. Even to get to the concessions often took weeks of arduous river travel, crossing uncharted land, and encountering a mosaic of local ethnicities and political alliances. At best, the concessions were usually

"occupied" by a handful of white men, the climate being too insalubrious, and the work too unpleasant, for European wives or families.

As the historian Catherine Coquery-Vidrovitch has pointed out
in her magisterial history of the concessionary system, in 1900 equatorial Africa was still a colony that no one had ever taken possession of, remaining largely untouched by Europeans. By the eve of
the First World War, the white population had more than doubled,
thanks to the influx brought by the concessions, but it still amounted
to only two thousand Europeans, in a possession four times larger
than France. The white men—and they were overwhelmingly men
(in 1900 only *seven* of the 248 white inhabitants of Brazzaville were
women)—who dotted the countryside were merchants, administrators, and missionaries. They felt constantly under siege, socially and
physically isolated, and short on cash and resources. And indeed, in a
few short years most of the concessions faced financial ruin, leaving
the government without a means of developing the colony.

Where companies survived and thrived, it was not the result of
fair and reputable business practices. Company agents were encouraged to collect the largest amount of resources in the least time and
at the lowest price. Colonial administrators helped concessionary
companies develop their industries by recruiting workers, organizing porters, and quelling any resistance. The most effective tool that
both agents and administrators could use was armed militias made
up of African soldiers, usually not of the same ethnicities predominant in the region in which they worked. Isolated white men living
in difficult conditions, under intense pressure to produce, aided by
well-armed soldiers, and interacting with communities many considered "savage" had predictably unfortunate consequences.

The concessionary system soon became associated with violence, including imprisonment, torture, rape, and murder. Wild
rubber, the most immediately profitable resource, was notoriously
difficult to harvest. Wild rubber plants had to be found, sliced, and

sapped, often in areas far from any villages. As a result, workers had to roam far and wide to collect their quotas, often being gone from home for days at a time. Villagers were motivated to harvest rubber by the imposition of a tax system designed both to raise money for the colony and to give the indigenous population what one official called "a taste for work."

In theory, agents and administrators went to local chiefs both to collect taxes and to find workers. They discussed the money or labor they wanted, relying on the authority of the chiefs to send men and women to work. Not surprisingly, the population steadfastly resisted taxation, refusing to pay, and if forced to work, they did so at a pace that did not produce the quotas the agents wanted. Left essentially on their own, administrators and agents soon devised ways to motivate villagers. To increase productivity, they created regimes of terror in which soldiers entered villages, abused and kidnapped women and children, and demanded that the able-bodied men collect a set number of kilos of rubber. When the kilos were delivered, the hostages—if still alive—were returned. This systematic intimidation often entailed beatings, rape, murder, starvation, and the destruction of property. Increasingly, entire communities fled the soldiers, depopulating the countryside. The *commissaire-général*'s office repeatedly reminded administrators in the field that taking hostages in order to collect taxes or force work was not acceptable. And yet reports continued to roll in that villages were being burned and fields razed.

Administrators were clearly involved in the intimidation of local communities, especially in Middle Congo and Ubangi-Shari. In mid-February 1905 a story hit the Parisian press that caused a scandal. Under the headline EXECUTIONERS OF BLACKS: A COLONIAL CRIME, the Parisian daily *Le Matin* reported the arrest of two administrators, Emile-Eugène-Georges Toqué and Fernand-Léopold Gaud, implicated in the deaths of multiple Africans. The most appalling incidents reported occurred on July 14, 1903—a year and a half before the story appeared in the press—during a celebration of the French national holiday. At a party of white men near Brazzaville, amid plentiful eating and excessive drinking, someone suggested "an entertainment as new as it was ingenious, which would justly mark this day of coming together." The Frenchmen decided to experiment with "the effect of dynamite on a negro."

This "brilliant idea" reportedly met with laughter and enthusiasm. The group captured a young African man on the square in front of the residence and forced him into an interior courtyard. He struggled in vain to free himself from the ropes that bound him. In a sign that the cruelty was thoughtfully carried out, the original intention of attaching the stick of dynamite between the man's scapulae was reconsidered. Instead, someone produced a copper tube that was used as a cannula to insert the dynamite in the man's anus. "A detonation rang out," *Le Matin* recounted to its readers, and "bloody debris, limbs, intestines" were hurled in every direction. Amused by what they'd done, the men plotted another atrocity. Sometime later they decapitated a second African man and boiled his head to make a dish they intended to serve to his friends and family.

News of Toqué and Gaud could not have broken at a worse time for defenders of the French Empire. Still fading from memory was a hideous scandal involving two other French officers, Paul Voulet and Julien Chanoine, who in 1898 had led a military expedition ostensibly to pacify regions of French West Africa. The column soon devolved into depravity, leaving charred villages and dead men, women, and children in its wake. Voulet and Chanoine were ultimately killed, but not before murdering a superior officer and badly scarring the reputation of the military and France's supposed "civilizing mission."

More immediately relevant in the French Congo were the extraordinary accounts of the brutality of rubber harvesting that had been emerging from Leopold's Congo Free State. In the months before the Toqué and Gaud story hit the stands in France, an international movement driven by writers, missionaries, and businessmen had publicized accounts of kidnapping, rape, and widespread killing in Leopold's Congo. The Congo Reform Association, founded in 1904 by Edmund D. Morel, Roger Casement, and others, sponsored lectures that provided accounts and projected images of men, women, and children who had been mutilated by Leopold's soldiers in their effort to motivate rubber harvesting.

With the European colony just to the south of the French Congo under international scrutiny, the French government set out to distinguish its approach to empire from Leopold's. But it would not be easy. The graphic details of Toqué and Gaud's deeds brought vivid images to mind; indeed, some French publications, like the satirical journal *L'Assiette au beurre,* left nothing to the imagination by printing drawings of ghoulish white men participating in various tortures of Africans. With newspapers across the political spectrum covering the story, politicians in the National Assembly had no choice but to demand an immediate inspection of the French colony. The goal was not to expose further horrors but to stem mounting criticisms. An investigation led by a known, reputable figure would insulate the possession from further criticism. And there was one obvious candidate.

———

PIERRE SAVROGNAN DE BRAZZA was called out of retirement—he had been living a quiet life with his family in Algiers—to head an investigatory commission. He agreed on the condition that he could choose his companions, a group that ultimately included several colonial inspectors, an official from the foreign ministry, and a young philosopher and outspoken socialist named Félicien Challaye.

Brazza on the cover of *Le Petit Journal*, March 19, 1905.

Challaye was the only member of the mission who was allowed to cover the process for the press, his dispatches appearing in *Le Temps*. Once again Brazza was offered the opportunity to try to reform the part of the world that had made him famous. But this time, instead of trying to remake the French Congo to benefit humanity, he and his colleagues would investigate colonial rule's apparent penchant for brutality and corruption.

The choice of Brazza reflected an official desire to reshape the negative image of French imperialism in equatorial Africa. The French press hailed Brazza once again as a hero who would restore humanity and peace to the region while also showing that broader comparisons to Leopold's Congo were unfounded. The triumphalist myths of late nineteenth-century imperialism produced an array of heroic images of Frenchmen abroad. The stock-in-trade of popular journals and commemorative publications was scenes of white men traversing great deserts, facing down marauding warriors, hunting vicious predators, and defeating slave traders and tyrants. Brazza made iconic a strikingly different vision of empire: that of the white man of peace, surrounded by curious and attentive villagers.

Fittingly, in March 1905, as his mission prepared to depart for the Congo, the widely read *Petit Journal* ran a color cover portraying Brazza in his younger *vaillant* guise, engaging thoughtfully with his escorts as villagers looked on excitedly. The arrival of this white man, the image made clear, was cherished and welcomed. Nothing could have been more at odds with the revelation of white officials torturing and murdering innocent people. On the flip side of Brazza's color image were articles that stated the challenges facing France in more detail. Blaming the scandalous acts of violence on an insufficient administration and on the difficulty of exposing "the very primitive natives" to "more civilized morals and to regular work," the newspaper said no one was better qualified to study the situation than Brazza. "The great explorer," *Le Petit Parisien* waxed, "still knows how, with a diplomacy of pure benevolence and humanity, to win the sympathy of the great African chiefs." It was, after all, the way Brazza had won so much territory for the empire "without having fired a single shot."

The Brazza mission brought with it a reputation for strict methods, attention to detail, and accuracy. Upon their arrival in May 1905, Brazza and his men started investigating allegations of widespread brutality, conducting interviews, visiting posts, and reviewing reports. The commission produced an extraordinary body of evidence, including a dozen reports of more than a hundred pages each. Unlike the British envoy Roger Casement's 1904 investigation of atrocities in the Congo Free State, however, Brazza's mission heard but did not lend much credence to the testimony of Africans, whom Europeans deemed unreliable and lacking in fundamental understanding of truth.

Much to the dismay of those who wanted Brazza to restore faith in the colony, the commission's discoveries implicated every corner of white society. It found the colonial administration to be complicit in promoting brutality, especially through the nonpunishment of known abusers, and in burying stories that might attract negative attention. Brazza exposed Émile Gentil, the ranking administrator in the French Congo, as not only unhelpful to his mission but also as a brute. Investigations uncovered evidence that Gentil committed what Challaye called "veritable acts of cruelty." For example, Gentil was reported to have ordered one of his African soldiers to summarily execute a man accused of stealing bananas. On other occasions, indigenous witnesses corroborated claims that Gentil had ordered men and women accused of minor offenses to be tortured or whipped with the *chicotte,* the hippopotamus-hide whip that was regularly used to torment and scar African subordinates. A variety of white officers, however, insisted on Gentil's innocence and respectability, leading the Brazza commissioners to determine that they lacked sufficient evidence to reach a verdict.

The mission found more convincing proof of regular collaboration between administrators, from Émile Gentil on down, and concessionary agents. Regulations prohibited administrators from intimidating or punishing villagers for failing to provide sufficient labor for concessionary companies. But there was evidence that they regularly did. Among other episodes, Brazza's commission found that agents of a concession near the post of Bangui collaborated with officials to motivate villagers to produce more rubber by kidnapping about forty women and some children. When their chief came to free them, the local administrator told him to "produce some rubber." The blunt reply suggests that everyone involved understood the colonial quid pro quo.

The women were then moved, along with others taken from neighboring areas, to Bangui. There the women and children—now counting 58 women and 10 children—were imprisoned in an airless hut that was 290 square feet (27 square meters), offering each woman and child about four square feet of space. The women, who'd been charged with no crime, were held in the cramped building for weeks. After five weeks, 45 of the 58 women and 2 of the 10 children had died. An inspector passing by chance through Bangui found the survivors, describing them as more emaciated than victims of any chronic malady. He was not asked to conduct autopsies; the bodies of the dead were dumped in a river. Gentil ordered an investigation, but low-level officials claimed they'd helped the women, whom they'd found abandoned in the bush. The story, which came to be called "les femmes de Bangui," was buried.

Brazza found that the kidnapping of women and children was common practice. While an administrative order had outlawed the practice in 1887, holding hostages had remained central to harvesting rubber and recruiting porters up until Brazza's mission and, indeed, beyond. In mid-July, while traveling in the region of Krébedjé, Brazza found 119 women and girls who had been taken hostage nearly two months earlier. An administrator in the region, knowing of Brazza's arrival, tried to hide the women, dispersing them in different villages. Brazza was tipped off, however, when a local dance troupe performing for his arrival incorporated elements that suggested the pursuit and capture of people. Brazza demanded to see the post's daily journal and discovered entries related to the hostages who'd been taken after a conflict between a local concession and villagers in the region.

The investigation also discovered that some concessions opted for even blunter forms of brutality. A concession near M'Poko, for instance, committed widespread abuses, including the wholesale killing of nearly a thousand men, women, and children in an effort to subdue the region and intimidate villagers into producing rubber. Despite revelations about the company, after Brazza's departure, some forty Europeans with a private militia of four hundred African soldiers hunted down anyone not working to expectations. What followed were, according to one report, "veritable massacres of blacks." A court in Brazzaville established that some fifteen hundred indigenous people had been killed. Despite the "many murders" committed, a few black agents of the company were punished, and its white owners escaped any sanction.

Brazza's work in central Africa took a toll on his delicate physical health. Near the end of his nearly five-month tour, he fell ill with dysentery, fever, and exhaustion and took to his bed. His team completed the field investigations, regularly delivering information to the convalescing Brazza for him to incorporate into his final report. Brazza spent all his energy thinking about the problems of the Congo. "When he had the energy to speak," Challaye later remembered, "he spoke of the Congo." Frail as he was, Brazza still called his compatriots around him to debate the most effective ways to deal with complicated issues like porterage, and how to protect Africans from the "tyranny" of the concessionary companies and the administration. In early September 1905, on his return to France, Brazza grew even weaker, finally disembarking at Dakar in hopes of recovering, but he died shortly thereafter.

His report was delivered to the Colonial Ministry after his death. Written in a sober and legalistic language, with cold details of corruption and atrocity, Brazza's critiques were clear and wide ranging. He found the French presence in the colony entirely insufficient; "it is superficial, incomplete, and powerless to pursue the civilizing goals that France sets in all its new possessions." Likewise, the military was poorly organized; the judicial system was lamentable (overseen by provisional judges who knew nothing of the law); and the whole colony needed better supply and communication lines, including railroads. Considering his commission's findings, it is not surprising that rumors abounded that Brazza had been poisoned— an idea even his wife believed—to cover up the atrocities and corruption. Brazza had shown no fear in condemning men who Gentil insisted were "apostles of development."

Upon Brazza's death, his commission was transferred to Jean-Marie de Lanessan, a former governor-general of Indochina and navy minister. Responding to concerns about potentially damaging fallout, Lanessan refused to publish Brazza's report. In its place, he produced a heavily redacted version of Brazza's findings that deflected blame from the colonial administration, including Émile Gentil, as well as from ministers in Paris. Brazza's evidence would see Gentil recalled from the Congo and eventually replaced, but he went unpunished for his corruption and acts of violence. Lanessan did not entirely exonerate the concessionary companies, but he found fault primarily with a few misguided and immoral actors—people like Gaud and Toqué—not with the colony's French administration.

A number of individuals and organizations concerned with protecting indigenous interests lobbied for the publication of Brazza's full findings, but without success. In the years just after his death—and for decades thereafter—Brazza's report was said to have been lost, destroyed, or misplaced. As time passed, it seems, few looked for it. It was not "rediscovered" until 1965, tucked away in the archives.

———

EVEN AS BRAZZA'S REPORT was buried, the administration of the French Congo changed in the wake of the hero's death. With Gentil recalled to Paris and replaced, a new administration took over, better funded (though still modestly) and with more offices to fill. A number of decrees clarified and integrated the administrative hierarchy in the region. Between 1906 and 1910, the name of the possession changed from the French Congo and Dependencies to French Equatorial Africa, and it became a single administrative entity that could have made sense only in the minds of colonial bureaucrats. The colony, which included Middle Congo, Gabon, Ubangi-Shari, and Chad, reached from equatorial rain forests to sun-scorched deserts, encompassing millions of people, dozens of ethnic groups and languages, and myriad religious traditions.

The number of administrators, which in 1900 was fewer than a hundred, grew exponentially in the wake of the Brazza mission to over five hundred. Outside the main towns, though, posts were still often staffed, at best, by a single white administrator, and high rates of attrition meant that when positions were vacated, they were not necessarily refilled. The new administration, still woefully inadequate for governing a massive and varied territory, at least represented an effort to move on from the concessionary system and to create infrastructure and industry through government initiative.

A railroad again became a centerpiece of a vision of a more prosperous, equitable, and humane colony. Indeed, the need for it was ever more pressing when copper, as well as tin and zinc, were identified as potentially lucrative commodities. French speculators were not the first to discover copper in the region—local communities had been producing copper goods for generations—but nearby Belgian interests had shown it to be profitable. In 1905 a French company made Mindouli, about 135 kilometers due west of Brazzaville, the center of its copper production in Middle Congo. Located on a potential route to the sea, and with mounting concerns about relying on the Belgian railroad (which was shipping Belgian-produced copper), Mindouli's potential riches brought engineers back to the question of how to build a French line.

Once again, speculators and engineers pushed new plans for a rail link to the sea, showing the same self-assurance as Léon Jacob had, even when their methods were remarkably haphazard. Marc Bel, an engineer of the mining company in Mindouli, deemed older plans along the Kouilou-Niari River to be no longer viable. Preferable, he argued, would be a line that crossed from Brazzaville to Mindouli and then on across to Pointe-Noire, some 20 kilometers south of earlier plans to locate the terminus near Loango. Bel possessed no specific information on the viability of his route, which would cross the treacherous Mayombe, and no detailed evidence of the desirability of Pointe-Noire as a port. He nonetheless convinced politicians in France to fund studies of his suggested path. He headed out on his mission of investigation in fine colonial style with another engineer, an army captain, four master miners, eighteen African workers, an unspecified number of porters, and his wife, Mme. Bel, a naturalist. While clearly working on behalf of mining interests, Bel nonetheless insisted that the train would be for the public good.

As speculative as Bel's ideas may have seemed, they rekindled interest in a railway and pushed forward funding of more serious planning. The most notable idea to emerge was the use of Pointe-Noire as the site of a port. A hydrographic study found Pointe-Noire—for centuries an assembly point for captives brought from the interior to be sold to European slavers—to have navigable sand-bars and currents, as well as an amenable climate for a deepwater port. Meanwhile the Société de construction des Batignolles, the third largest public works firm in France, internationally renowned in metallurgy and mechanical engineering, was commissioned to conduct topographic studies for possible routes from the coast through the Mayombe and on to Brazzaville.

The route originally outlined by the Batignolles looked relatively straightforward on paper. The 540-kilometer route would progress 70 kilometers east of Pointe-Noire through easy, hilly country. The following 100 kilometers would traverse the more mountainous Mayombe, though the engineers did not foresee climbing to an altitude higher than 350 meters or needing to build more than eight relatively short tunnels. From the Mayombe, the rails would even out across the Yangala savanna and the Niari-Moyen plain for some 175 kilometers. The initial proposal called for a fairly difficult final 200 kilometers across the rocky, hilly terrain before reaching the sandy massif of the banks of the Congo. With a one-meter-wide rail, the Batignolles proposed a train that could haul 120 tons each direction at a less-than breakneck speed of 25 kilometers an hour. The price tag was estimated at 95 million francs, which adjusted for inflation would be more than US$400 million today.

The Batignolles plan was the most promising to date; it looked like the colony would finally break ground on a railroad. In July 1914 Governor-General Martial Merlin approved the plan and received a loan to start construction. Three weeks later much of Europe was at war and construction was put on hold. Far from the trenches of the western front, engineers and business interests in the Congo spent the First World War debating more possible routes as well as the merits of those already proposed. When people discussed the railroad, as one colonial paper put it, "a wave of skepticism and also of discouragement invaded the colony." But if agreement about details was in short supply, the assumed need to build a railroad at least remained consistent.

Once the war ended, a renewed French interest in the empire, as well as the emergence of new ideas about economic development, gave many hope that the railroad's moment had finally arrived. To many Frenchmen, the empire's inhabitants—especially in Africa— were trapped in stagnant societies plagued by superstition, cannibal- ism, and tyrannical political practices. Colonial reformers thought in terms of what their subjects *lacked*: knowledge, political stabil- ity, and—crucially—technology. They saw infrastructure as key to enriching local economies and to solidifying political control. Roads, waterways, and railroads had transformed nineteenth- century European states; now supporters of empire argued that sim- ilar projects would pull colonial populations out of the prehistoric mire into which their cultures had allegedly sunk.

No one made the case for colonial modernization—including the importance of railroads, ports, and other infrastructure—more forcefully than Albert Sarraut, who, as colonial minister from 1920 to 1924, wrote and spoke extensively about the *mise-en-valeur*, or development, of France's empire. While certainly not all of Sarraut's ideas became the law of the land, his eloquent vision of empire influ- enced many pro-colonial lobbyists. Never doubting that colonial- ism demonstrated French greatness, Sarraut nonetheless insisted that it ultimately benefited humanity. Sarraut's rhetoric updated Braz- za's formulations for French Equatorial Africa. Dismissing colonial oppression as a thing of the past, Sarraut called for a "harmonious accord of races" as well as collaboration between Frenchmen and local populations to forge new societies together.

For all its supposed generosity, Sarraut's approach aimed in no small part to lessen the empire's financial drain on the French bud- get, a growing political concern in the wake of the First World War. Across the political spectrum, French politicians were eager to grow the nation's economy, as the center-right statesman Raymond Poincaré put it, by "developing our magnificent colonial domain." Imperialism was a logical extension of capitalist investment, at a moment when France was trying to rebuild after a devastating war. But profitable aspirations did not minimize the plan's selfless aims. In Sarraut's mind, colonialism, rather than serving national interests, forged "human solidarity," which was in fact its "only source of legitimacy."

Sarraut echoed many of Brazza's formulations, but he was also a product of an era of globalization where the rhetoric of national triumph in Europe's empires was increasingly tempered by allusions to shared humanity. Advocates of empire in the 1920s preferred the language of trusteeship and philanthropy to that of colonizer and ruler. Sarraut contributed to this new discourse. "In the name of humanity's right to life," he said in a 1923 speech, "colonization, agent of civilization, will take charge of development"; it would help those who possessed resources but lacked the knowledge to bring them to market. "It is for the good of everyone that we act," Sarraut concluded, "and primarily, for the good even of those that appear dispossessed." Colonial *mise-en-valeur* was to be, not about exploitation, but about helping colonial subjects help themselves.

Sarraut's claims that colonialism could serve—and enrich—all of humanity inspired and helped justify myriad projects around the empire. His rhetoric, with its promises that empire could spread peace, wealth, and happiness, was certainly mellifluous to the ears of many a colonial apologist. It mixed the idealism and selflessness of fallen heroes like Brazza with an institutional and economically grounded agenda for the future. Economic development would not only justify and fuel French power abroad, then; it would also provide a humanitarian narrative for empire building.

# CHAPTER 2

# THE RIGHT MAN FOR THE JOB

To awaken Sleeping Beauty's palace, could we choose a more dashing knight?

ALBERT LONDRES

AT THE END OF the First World War, the building of the railroad became a key component of the economic development of Equatorial Africa. The project also came to symbolize, at least rhetorically, a new era for the colony, a departure from the days of the concessionary system, with its abuses and unprofitable monopolies. It became emblematic of a rejection of the administrative apathy of the past. Albert Sarraut saw the railroad as an urgent necessity, a "question of life or death," for a colony that had to open its goods to a world market. Moreover it was part of a larger vision encompassing other programs as well, including efforts to end hunger, child mortality, and disease.

In 1920 Victor Augagneur, a radical-socialist politician with extensive colonial experience, became governor-general. With the white community losing faith in the likelihood that a railroad would ever be built, and the economy stalling, Augagneur pushed ahead aggressively. In late 1920 he sought and received authorization to begin construction on a short section of a rail line departing Brazza-ville. The rest of the route—that is, the five hundred or so kilometers beyond the outskirts of Brazzaville—remained unmapped, still the subject of controversy and disagreement. Augagneur reasoned that the first section, leading out of Brazzaville, would be the same regardless of the final route. Fittingly, the colonial minister who signed the early paperwork for the railroad loans to the Congo was Albert Sarraut.

At ten o'clock on the morning of February 6, 1921, in front of the entire white population of Brazzaville, Augagneur's wife had the honor of breaking ground, a task she performed in a white dress and a large straw hat. She was the first and likely the only white woman ever to move dirt on the construction site. Her husband, decked out in a white suit and pith helmet, then gave a speech. He admitted that some would consider it an odd moment to start construction. There was no established route for the railroad to follow. There was no port, or even agreed-upon site for a port. There was no policy that identified the thousands of men who would be needed to com-plete the construction. "We could seek the best gradients, the best curves, and there would still be complaints," Augagneur explained. "I say again: we must see and act." The railroad was under way. Nearly forty years after Brazza first suggested the need for a train, and after more than two decades of serious discussion of routes and specifications, Augagneur—as well as the colony and nation behind him—had found a way to act impetuously.

Augagneur's decision to start building a railroad unprepared, without a route, blueprint, or employees, was emblematic of the hubris that would drive the construction through to completion. One momentous problem he knew to expect was a shortage of labor. In 1919, more than a year before construction commenced, he acknowledged an "absolute shortage" of the skilled artisans and even the untrained laborers needed to build the railroad. He sent a representative to French West Africa to look into the possibility of recruiting workers there. But, not surprisingly, officials in Dakar were less than enthusiastic about the idea. Equatorial Africa would have to use its own workers.

The issue of who would build the railroad became all the more pressing in the summer of 1922, when Colonial Minister Sarraut signed an agreement with the Société de construction des Batignolles to begin construction of the line from Pointe-Noire to Kilometer 172 toward Brazzaville. It was a plum agreement for the Batignolles, which was essentially guaranteed a 20 percent profit above the projected cost of the railroad and which had only to oversee the workforce on the construction site. And perhaps best of all, the company did not have to find its own labor. The recruitment, housing, and feeding of workers was left to the colony, to the governor-general's office, and to administrators in the field.

For the colonial state, no single dimension of the railroad was more crucial—or more difficult—to achieve than recruitment. African bodies would be the primary tools of the building. African hands, arms, and backs would cut trees and clear bush, hoist rail and cement, and literally move mountains, one rock at a time, often without the benefit of steel implements. By agreement, the government would provide the Batignolles with a constant stream of men. It would be no small feat: between 2,000 and 8,000 workers would be needed at any one time to keep construction on schedule.

Labor was initially recruited from the area directly adjacent to the projected railway line. There was much logic to the decision. Workers could in some cases live in their home villages, or at least regularly visit, where they could get supplemental food and other supplies. A study done before the First World War had suggested that men from the region would be far better suited to the climate than those brought from afar. Citing experiences in French West Africa, one engineer determined that exposing foreigners to such a climate would be "an <u>ineffective and even barbarous</u> means" of using labor. He felt so strongly about it, he underlined the judgment in his letter to the colonial minister.

In practice, however, relying on local labor quickly faced major drawbacks. The population near the projected route was small, providing few workers in the first place. Making matters worse, men living near the route increasingly resisted the call to recruitment. In the spring of 1924, one administrator reported that the 301 recruits he had brought to Brazzaville were found only with difficulty. The cause of his troubles? "The rumors circulating in the country." Workers spread the word that they were subjected to "bad treatment" on the worksite, including "no weekly rest day, abusive withholding of salaries, beatings, etc."

The rumors were based in fact. The resulting resistance led recruiters to use soldiers to force men to "volunteer," much to the dismay of some officials. Augagneur told his men that it was "by persuasion and not by coercion that you will drive the natives to work." But persuasion would be effective, one official would later point out, only if the administration could guarantee the security and livelihood of all laborers.

No such guarantees could be made. In May 1924 a health inspector named Dr. Levet discovered that workers were deprived of basic necessities. The doctor reported that, during his twenty days of touring, he witnessed workers in "a state of lamentable physical decay" that bordered on "starvation." Levet's was but an early voice to warn of the failures of official rations. "One certain fact" was that recruits did not get the required 500 grams of rice promised them. And even if they did, for a man who worked five hours before lunch and five hours after, "the energy value" of such a ration—about 650 calories—was "manifestly insufficient." Chronic hunger on the construction site should have dissuaded officials from bringing more men and women to the railroad, but it did not.

Morale plummeted in the region, while desertions of recruits rose. In Mouyondzi, a village near the line, a "real fear" of being forced to work on the railroad led to political instability and even violence. Other villagers simply disappeared into the forest, often crossing the river into the Congo Free State to avoid recruitment. The resulting exodus caused local food production—a major source of food for workers—to drop off precipitously. As the construction moved away from the ends of the line, more and more men were needed for clearing the bush and starting the construction; they were needed for porterage, to carry in food and building supplies. The potential labor force near the construction site was simply not sufficient in the long term.

Augagneur had been either bold or reckless, depending on one's perspective, in getting the railroad started. But years into the construction, problems still plagued the project. With the announcement of the French railroad, the Belgians undertook to improve their railway between Léopoldville and the port of Matadi, leading many French critics to wonder whether it was worth it to have their own. The pro-colonial French press, which had initially been cautiously optimistic about the project, started to second-guess it. For example, *La Dépêche coloniale,* an influential newspaper, questioned the logic of building a railroad in "a wretched country" when doing so would ruin "an already too rare labor force." Augagneur was increasingly criticized for his indecision and poor judgment. As late as 1924, there still was no definitive route planned through the mountainous Mayombe. One newspaper claimed that even the Batignolles were doubtful about piercing a landscape of "quartzose and granite crags," as well as digging a port at Pointe-Noire. But abandoning the railroad was not an option. With criticism mounting, the French administration reacted in tried-and-true fashion: Augagneur had to go.

IF PIERRE SAVORGNAN DE Brazza was the first to conceive of a railroad in the French Congo, it was another Frenchman of Italian ancestry, Governor-General Raphaël Antonetti, who built it. Antonetti's decade-long stint, from 1924 to 1934, as the head of Equatorial Africa would coincide almost exactly with the construction of the railroad. He greatly expanded the effort upon his arrival, and his speech at the inauguration of the railway in 1934 was one of his last official public presentations before retirement. While from the state's perspective Antonetti certainly "oversaw" the project, the term fails to capture the relentless determination he showed as its chief coordinator and its most vocal defender. His obsession with it transformed the project, at least in his own mind, into what he insisted was "one of the world's great railroads." It was no small claim for a line shorter than the distance from Paris to Lyon, or San Francisco to Los Angeles.

Nonetheless, his commitment to the project became known far and wide; Antonetti became, as one publication dubbed him, "the Governor of the Congo-Océan." Despite his imperious name and his ascension to the highest post in French Equatorial Africa, Raphaël Valentin Marius Antonetti was unremarkable in many ways. Perhaps fittingly, the man who spent much of his tenure as governor-general dodging and denying claims made against his administration left little evidence of his own history behind. He hid his personal life from view. Almost nothing is known of his early life; indeed, he seemed to have been particularly opaque about his youth. He was born in Marseille in 1872; some have suggested he was Jewish, but even that remains unclear. Some sources say his family was Maltese, though a plaque at the hotel-restaurant in Pointe-Noire that today bears his name claims they were Corsican.

Upon his death in 1938, only a few mainstream French newspa-

Governor-General Raphaël Antonetti in his Paris apartment, 1932.

pers even noted the retired governor-general's passing. *L'Homme libre,* the daily founded by Georges Clemenceau, buried a two-sentence obituary on page three, noting only the barest facts of his life. The Catholic *La Croix,* one of the only national papers to run the news of Antonetti's passing on the front page, offered platitudes: he was a "great administrator" remembered for "magnificent achievements," including the Congo-Océan. The *Journal des débats* wrote a bit more, not all of it accurate, but nonetheless tucked his obituary alongside the death notices of a senator's daughter and a former cook of Emperor Napoleon III. At his death, little could be recounted of him beyond a list of the dates and locations of his posts as a colonial bureaucrat. If he had close friends, none left fond remembrances of him. In the decades since, even major reference guides to the history of the French Empire don't always get his life story straight.

A photograph of Antonetti taken on New Year's Eve in 1932, a couple of years before the completion of the railway, reveals something of his personal qualities. He stands impeccably attired as a paragon of the accomplished functionary—simple, serious, humorless—in front of a marble mantle in his apartment in Paris, where he was on home leave. The photo shows him as a man of cold certainty, staring into the distance, holding his tightly clenched fist before him. This contrived gesture of determination was typical of an era when self-confidence in France's imperial role of *civilisateur* was regularly on show.

Unintentionally, however, the photo also betrays the emptiness of the promises of empire. Antonetti was a collector of African art—part of his collection is now housed at the Musée du Quai Branly–Jacques Chirac in Paris, France's premier museum of non-European art—but in the photo, the carvings that decorate his mantelpiece seem out of place next to this man who spent decades in Africa. Next to a more dynamic personality, the elaborate, stylized carvings, obscured in the dark tones of the photo, could have suggested an active imagination or emotional complexity. But placed as they are next to Antonetti, the luster and expressive lines of their dark wood contrast with his stolid visage. His indistinct pallor reflects the bureaucratic ethos often at the heart of modern empires—an ethos distinctly lacking charisma, creativity, and warmth.

It is tempting to imagine that the administrator who marshaled the building of a railway through difficult topical terrain, who coordinated the movements of tens of thousands of men, who showed unwavering sangfroid in the face of many thousands of deaths, possessed great, if troubling, brilliance. But we have little reason to think Antonetti was that complex. He lacked depth or even much curiosity. Unlike many colonial officers who tried their hands—with mixed success—at writing ethnographies, local histories, or studies of colonial rule, Antonetti betrayed no scholarly aspirations. And

he certainly never appeared morally torn about, or even apparently
aware of, the ironies of empire. Instead, he admitted the existence
of no contradiction between the rousing promises and the horrific
realities of colonial development projects. He never questioned his
own motivations or the necessity of the methods he employed. Any
suggestion of wrongdoing was summarily dismissed—both in his
mind and on the record. A lifelong bureaucrat, he was as much a
product of imperialist ideas and actions as he was a producer of them.

———

TO UNDERSTAND WHAT MADE Antonetti the ideal man to oversee
the construction of a railroad in the Congo, we must begin with
his training. A career colonial bureaucrat, Antonetti knew no professional life other than an administrative one, entering the colonial
service at age nineteen. He received no advanced academic or professional training; the École coloniale, the institution that prepared
many men for careers abroad, had only just opened when Antonetti
left France. The possibility of advancement no doubt appealed to the
young Antonetti. He started at the bottom, as an *écrivain,* a job that
consisted of copying other people's correspondence. His early years
followed a circuit that somehow took him to the Society Islands and
to the Comoros, two archipelagoes separated by over 16,000 kilometers (nearly 10,000 miles) of ocean. Within a few years, he was promoted to the secretariat-general, consisting of colonial paper-pushers
who were only slightly more respected than *écrivains.* But from there,
his trade became a profession, and his career moved steadily forward.

With the exception of a two-year stint in Saint-Pierre-et-
Miquelon, the windswept French possession in the northwestern
Atlantic known at the time mainly for its failing fisheries, Antonetti
spent most of his career in Africa. In 1905 he became secretary-
general in Djibouti, the port capital of tiny French Somaliland, a
scorched outpost that many considered a "hellish land." After four

years there, he started a fifteen-year stint in French West Africa, where he bounced from one appointment to another. He served short spells across West Africa, including in Dahomey (present-day Benin), Senegal, and Niger. In 1917 he finally settled down in Ivory Coast, where he became lieutenant governor and then governor. It was from this position that he was promoted in 1924 to governor-general of French Equatorial Africa, a post he retained until his retirement from the colonial service in 1934.

Antonetti was undoubtedly capable; he moved competently up the career ladder. That said, the bar for success in the colonial administration was low. Raw talent and intelligence were not essential: of the men who made up the ranks of the French secretariat-general in Africa, the Colonial Ministry itself deemed over two-thirds to be unqualified for their jobs. As late as 1913, many applicants to the colonial bureaucracy in West Africa failed basic dictation and arithmetic tests—but were still hired. In more colorful if uncharitable terms, one settler described administrations abroad as "depositories of excrement," where the lowest of society's low were flushed. Even Georges Hardy, a great commentator on colonialism and onetime director of the École coloniale, admitted that when a man joined the colonial service, people at home wondered, "What crime must he have committed? From what corpse is he fleeing?"

If Antonetti was fleeing a corpse, he hid it well. He likely shined amid the incompetence around him, being a perfectly literate and, at times, even powerful rapporteur. Coming of age in the colonial bureaucracy deeply informed his later work as governor-general of Equatorial Africa. His understanding of colonial rule, his presumptions about the limitations of Africans, and his abiding faith in the transformative possibilities of European technology all bore the hallmarks of a man of limited and often patently racist vision. For someone who traveled the world, he remained remarkably provincial in his views on how best to colonize foreign societies. In this way, he was not unique but rather shared the perspective of many French officials in Africa. Knowledge of local people and cultures and an awareness of the personal impact of colonial policies were not cherished traits of colonial officials. Despite living some twenty-five years on the continent, Antonetti appears not to have spoken a word of any African language.

Antonetti's early career saw him locked deep within the workings of the colonial bureaucracy. As secretary-general in Djibouti, he likely did little other than paperwork, sitting behind a desk copying out reports and letters, organizing and filing correspondence in the heat. Located next to Ethiopia and British Somaliland, at the confluence of the Gulf of Aden and the Red Sea, the port was a fast-growing center of international commerce. It had a remarkably diverse, if small and transitory, population. Indians, Egyptians, Greeks, Ethiopians, Yemeni Arabs, and Jews all worked in various aspects of the economy, as anything from merchants to day laborers. Antonetti no doubt had daily interactions with some of them. But he spent the vast bulk of his waking hours indoors with like-minded Frenchmen engaged in the monotony of paperwork.

As Antonetti's career moved to West Africa, he had more opportunities to interact with Africans, though his engagement was still limited and, increasingly, begrudging. Before the 1920s, the administration in West Africa had little interest in seeking the opinions of subjects, even of local elites. Rather than mingle with their subjects, lieutenants-general relied on officers in the field—*commandants de cercle*—to send updates of what was actually happening in the countryside. Men in Antonetti's position, then, spent much of their time receiving news from subaltern French administrators, as well as from local French business interests and settlers, and filtering the information into reports that they sent on to their French superiors. Notably left out of this chain of communication were African perspectives.

A product of this institutional culture, Antonetti was never one to consult Africans if it could be avoided. Time and again he argued that Africans could at best complicate, and at worst endanger, French rule. In 1914, acting as lieutenant governor of Senegal, Antonetti was outraged at the recent election of Blaise Diagne, the first African deputy to have a seat in the National Assembly in Paris. Diagne, who represented the small number of Africans who possessed French citizenship in Senegal, had been outwardly critical of the government's ineffective response to a plague epidemic ravaging the colony. Antonetti wrote to the governor-general, incensed by all that Diagne represented—that is, African involvement in colonial affairs. Overreacting to the amount of real power Africans had, Antonetti went on a tirade, suggesting colonial affairs would soon be determined "by our porters, our copyists, by fishermen, by people of another race, ignorant, fanatical, 90 percent illiterate." Antonetti insisted that Diagne's election today would lead to "blacks" taking over the police tomorrow: "and they will be able to use the armed force against us." He implored the governor-general to nullify Diagne's victory and, if that didn't work, Antonetti suggested, as if channeling a gangster, to "get rid of him." The governor-general, William Ponty, though apparently sympathetic, thought such actions would be "grossly immoral" and wisely omitted the suggestion in his report to Paris.

Antonetti showed no greater interest in local attitudes when he was promoted and posted to Ivory Coast. There he argued *against* a policy calling for greater engagement with local chiefs. He also earned a reputation for brutality. During the First World War, he put in place harsh methods of recruiting unwilling men for military service, justifying his approach by citing the need to support the war effort. One historian has described Antonetti's policy as "the sad alliance of misery, the pressuring of chiefs, and administrative coercion." In 1919, now dealing with an intransigent workforce, Antonetti wrote that the "fiercely individualist" local people obeyed only when "under duress." He oversaw a system of forced agricultural production, price fixing, and strict taxation of a subsistence population—all while championing what he deemed, without a shred of irony, the "generous goals of France's civilizing mission." All that really mattered was that the colony met its economic goals.

Antonetti's experiences in West Africa left an indelible mark on his ideas about colonial rule. He was not a man of nuance and was prone to wide-sweeping generalizations. He made, for example, few distinctions between Ivory Coast and the Congo, despite the colonies being separated by some four thousand kilometers, not to mention having vast linguistic, historical, and cultural differences. In a short pamphlet written in 1921 while lieutenant governor, he described Ivory Coast as inhabited by murderous forest people who attacked caravans looking for merchandise and human flesh. The forests of West Africa, he continued, were peopled by "a race that poverty and privation have driven to the greatest degree of savagery and physiological misery." More than a decade later, as governor-general in Equatorial Africa, he described the people of Congo in almost identical terms: "the populations have always lived miserably as victims of deprivation, of internal wars, of cannibalism, of the horror of ritual customs—of looting sultans and slave traders."

Antonetti expressed remarkably simplistic and uncharitable opinions of "the Africans" and the land they inhabited. For him, Africa was a collection of stagnant societies, trapped by savage customs and sadly lacking in European technology and organizational know-how. This was an attitude widespread in colonial administrations of the time, and a vision of empire that Antonetti came to embrace as a bureaucrat. Civilization was France's gift to its colonial subjects, but French administrators also had no qualms about forcing it on local populations, regardless of the suffering caused. What made Antonetti remarkable, though certainly not unique in the empire, was the intensity with which he espoused his beliefs and pursued their implementation.

ANTONETTI'S GOALS FOR COLONIAL development corresponded well to attitudes that emerged in the wake of the First World War, when *mise-en-valeur* ruled the day. While still in West Africa, he espoused support for grand projects, such as infrastructure development, campaigns to eradicate disease, and education initiatives. Pursuit of such projects both addressed the financial needs of colonial investors and allowed the French to show what Antonetti called "our solicitude for the black race." In a published pamphlet, he outlined in stark terms the radical—and seemingly immediate—effect that building a railroad in Ivory Coast would have on "millions" of people. Written for a series called "Colonial Problems" published by the Committee for French Africa, Antonetti's short article explained how a simple railroad line would transform all of West Africa by linking regions that were long isolated from one another but possessed complementary features.

According to Antonetti, the trade routes across the Sahel and savannah of the French Sudan (modern-day Mali, not to be confused with present-day Sudan or South Sudan) had historically been cut off from the sea by the forests of northern Ivory Coast and the Gold Coast. French Sudan was not simply a string of trading posts, he argued; it was a land rich in history and fable, where salt, shea butter, rugs, and captives were moved and traded. With an unexpected splash of orientalist imagery, Antonetti explained that "grains, local textiles were bought by the Moors and the Tuaregs; slaves, gold dust, ivory by the Arabs and Moroccans, who came to Timbuktu from Tripoli, from Tunis, from Algiers, from Morocco. It was the Mossi who provided the black eunuchs to the harems of Constantinople." French conquest had added to this trade by improving transport with the building of a railway between Kayes and Koulikoro in the French Sudan. But the sea remained difficult to reach. It required crossing either thousands of miles of desert to

Niger or the forest to the south, which, he claimed, was "an object of dread" to everyone in the region.

Here French ingenuity and industry entered into Antonetti's story: the "great program" to build a railway through the forests to the ocean would now "change the economic life of several million men." By linking Ivory Coast with the French Sudan, the coastal region would become integrated into the markets to the north, while the savannahs of the interior would find some 4 million new producers and consumers to the south. Antonetti stressed that the railroad would bring one region that was "mediocrely fertile but well populated" (the French Sudan) into contact with an area "extraordinarily rich but little populated" (Ivory Coast): *"Au Nord les bras, au Sud les richesses."*

The transformation Antonetti proposed was not just one of better economic trade; rather, it promised to remake African society in almost utopian ways. Workers from the north would seek fortunes in the underdeveloped but fertile south. With them would come cattle, whose meat would be sold to the underfed population of Ivory Coast. This was not an issue of better diet; it was the answer to all that ailed sub-Saharan Africans. Hunger had long rendered these people, in Antonetti's words, "very lazy and incapable of efforts." Better food, especially meat, would give them the chance "to fight against the physiological misery" that had long decimated the population. It all seemed so simple: to turn hungry, lazy Africans into productive, healthy workers, all that was needed was a railroad.

Antonetti's experience promoting railroad construction made him an obvious choice to replace Governor-General Augagneur in French Equatorial Africa. Much like the reorganization of the colony after Brazza and again after the First World War, Antonetti's appointment promised that the colony would be run in a rational, profitable, and humane fashion. On a publicity junket in France before his departure to Brazzaville, Antonetti freely espoused the language of humanity and economic development. A railway, he argued, would open access to copper, rubber, lumber, and other resources of great value. This, he insisted, could be done with the willful help of African laborers and without denying them "the French duty of humanity." Even the skeptical *Dépêche coloniale* seemed mollified by Antonetti's determination. On the afternoon of his departure from France, it printed an optimistic, if cautious, assessment: "Today a new era opens for the Congo; it ceases being the pariah of our colonies."

Antonetti's work began in earnest upon his arrival in the colony. It was then that he gave the railway its definitive name—a sign that reinvigorating the project was his main priority. No longer would it be the Brazzaville-to-the-ocean railroad, the Brazzaville-Pointe-Noire railroad, or any of the other wordy, unmemorable monikers people had given it. It was now a grand undertaking to connect the entirety of central Africa to the outlet of the mighty Atlantic; it was the Congo-Océan.

When he finally brought his regional administrators together in Brazzaville (it took some of them weeks or months of travel), Antonetti delivered his message to the colony itself. He was not a born public speaker, but his rhetoric could be as lofty as that of many another governor-general. His sights were fixed on nothing short of the transformation of Equatorial Africa. "To open for free trade access to regions heretofore impenetrable," he waxed, "to provide agriculture with valuable labor by reducing the use of porters, to furnish business and the administration with educated native auxiliaries who will serve at the same time as role models for their backward sort, this is the first stage, the solid foundation that is essential to begin the development of the Colony."

The Congo-Océan would be the cornerstone of the foundation. He admitted that many questioned the usefulness of building a railroad at a moment when railroads seemed somewhat passé. It was not an issue Antonetti felt he needed to belabor; no one, after all, questioned the value of railroads in France. Yes, it would be expensive and would demand a "painful effort" of the indigenous population. But being able to travel easily from Brazzaville to Pointe-Noire in the middle of the rainy season would mark "a turning point in the economic history" of the colony. He promised that the railway would also eventually stretch north from Brazzaville to Bangui in Ubangi-Shari and continue on to Fort Lamy in Chad, creating the "dorsal spine" of French Equatorial Africa. Once in place, this new network of railroads would transform not only the economy but also the very existence of the indigenous population.

As in Ivory Coast, the Congo-Océan railroad was meant to solve the problems that threatened Africans the most: their own habits. Diseases and the natural elements were devastating, he admitted, but "the most serious scourge" afflicting the colony "is the laziness of many of its inhabitants, their lack of economic foresight, absolute even among the races often considered relatively evolved." When money could be made, he insisted, the African chose instead to party. When starvation loomed, the African spent all his savings foolishly. Giving voice to the racial hierarchy that thrived in the empire, Antonetti waxed: "Ah well, gentlemen, we have towards these big children a duty of guardianship that we must exercise to the fullest." Antonetti's vision of empire—from his faith in infrastructure to his prejudices about indigenous people—informed his plan of remaking Equatorial Africa at nearly every step.

But if the new governor-general's vision and rhetoric of what the Congo-Océan could do for Equatorial Africa were well honed, he still faced the very considerable practical challenges of building a railroad in a tropical region lacking lines of communication, administrators, and laborers. For all of Antonetti's promises about the future, in the present the railroad remained behind schedule and lacking manpower. With his trademark pluck, he ordered labor recruitment for the railroad to begin across vast swaths of the colony. Gabon's heavy reliance on lumber production was deemed too important to disrupt, so Antonetti pushed for recruitment from the forests of the Middle Congo to the edges of the Sahara Desert of Chad, more than a thousand miles north. With a stroke of his pen, the governor-general put in motion what would be one of the largest forced migrations in the empire since the end of the Atlantic slave trade.

Recruitment in Middle Congo, Ubangi-Shari, and Chad expanded greatly the potential pool of men and women, some of whom were little more than children, who could work on the railroad. As a policy, it remained true to the fashionable belief that all of France's subjects should contribute, in sweat and currency, to the development of the colonies. But as an implemented policy, recruitment would soon jeopardize the empire's other assumed tenets, especially its much-heralded commitment to humanity. Moving forward quickly with construction had not served Augagneur well; Antonetti's audacity in recruiting workers would serve him—and the men and women who were recruited—no better. The Congo-Océan's troubles had only just begun.

CHAPTER 3

———

# THE PACHA PRELUDE

Here beyond men's judgments all covenants were brittle.
CORMAC McCARTHY, *BLOOD MERIDIAN*

WORD SPREAD QUICKLY. FROM village to village and region to region, along branches of families and trade routes, translated from one language to another, tales of recruitment were passed along. Albert Londres, one of the most famous French journalists of his day, chronicled the reactions of would-be recruits to the news that colonial administrators were coming. "From the Congo River to the Sangha, from the Sangha to the Shari," he wrote, "all you ever hear talked about is the Batignolles!" No village, no group was left alone. "The Bakotas, the Bayas, the Linfonos, the Saras, the Bandas, the Lisangos, the Mabakas, the Zindès, the Loangos are all snatched from their thoughts and sent 'to the Batignolles'!" They understood what recruiters wanted—"the White Man was coming to look for men for the railroad"—and those who could fled into the bush. Villages were abandoned; districts were left empty. Recruiters were met with shouts of foreboding: *"La Machine!"* The machine had arrived, and the machine ate men.

The arrival of recruiters in regions across the colony was one of the most intense and potentially life-altering encounters that men, women, and children had with colonial rule. Recruitment had existential consequences for entire communities. It separated parents and children, siblings and cousins, friends and lovers. Those taken were lost to their communities for years and often forever. Gone with them was the work they did for their communities and the roles they played in families and other social organizations. As Londres put it, for men and women to work on the Congo-Océan was "to run toward death." For many communities, recruitment meant, if not actual death, total loss—temporary or permanent—of kin, warriors, hunters, homemakers, loved ones.

The methods used to recruit were the subject of much debate for critics and defenders of the Congo-Océan project as a whole. From the beginning, and for years to come, Antonetti insisted the workers were "volunteers"—a departure from concessionary practices as well as a nod to the free market ideals of modern colonial *mise-en-valeur.* Critics, however, derided Antonetti's term *volunteers,* asserting that recruits were actually obtained by the tried-and-true means of violent coercion that had not changed since the heyday of the concessionary system. Rather than volunteers, Londres saw Antonetti's recruits as forced laborers; as in times of old, he reported, "entire villages were *punished*" in the colony's search for workers. A host of other critics documented similar abuses.

Not surprisingly, Antonetti bluntly denied such claims, pointing to the highly regulated nature of the process. Ideally, recruitment was meant to be peaceful and orderly. Antonetti's office, in consultation with the Batignolles and other advisers, contacted local administrators in the field, prescribing a certain number of men for recruitment. Administrators then turned to local chiefs—of tribes, villages, or families—for assistance. The French saw chiefs as an essential part of the administrative chain of command; they negoti-

ated with French officials for the delivery of able-bodied men as well
as supplies for the construction site. When chiefs did not seem to be
allies of the French cause, administrators replaced them, reshuffling
local political networks to benefit colonization. Intimidation, kidnapping, and violence were strictly prohibited.

An early, particularly well-documented case of a recruiter in
the field, however, provides a snapshot of the challenges as well
as the brutality of efforts to recruit "volunteers." It also reveals
how the coercion and murder employed by concessionary companies continued to inform the tactics used by officials recruiting for
the Congo-Océan. The central figure in the episode was Georges
Pacha, a young and largely inexperienced administrator in a remote
and unsettled part of the colony, near the border of northern Middle
Congo and Ubangi-Shari.

Pacha's efforts in the field had little in common with the recruitment procedures prescribed in Brazzaville. In February 1925,
responding to Antonetti's order, Lieutenant Governor Jean March-
and wrote to an administrator in the far north of Middle Congo, a
Mr. Augias, ordering him to provide laborers. Marchand, who had
been in the colonial service since 1898, knew how to motivate his
subordinates. Having emphasized their common goal—"the construction of the railroad must come before all other interests"—he
tried to instill a sense of healthy competition in Augias, noting that
other regions in Middle Congo had already provided men. Appearing flexible, he told Augias that while he should demand four thousand recruits, he would allow him, this first year, to send just five
hundred men. And they had to come immediately.

From Brazzaville, recruitment seemed easy enough. Local
chiefs would provide recruits, who would then be examined so as
to eliminate those who were too weak or too sick. Healthy recruits
were to receive a stipend of 0.75 centimes per day to buy food and
other necessities on the long journey to Brazzaville. Some male

recruits were allowed to bring their wives, who could help clean and cook while their husbands worked, though Marchand warned that women would find the transport to the railway arduous. Under no circumstances were children to be brought. Local administrators also needed to find guards—"serious and vigilant"—to accompany the recruits to Brazzaville. Two riverboats made the trip regularly and were ready to take human cargo. On paper, the process could not have been more straightforward.

But at the other end of the correspondence, the process was daunting. The response from Augias made clear how unprepared officials were to recruit villagers. In painfully deferential terms, he replied to his superior that he "thought it necessary to respectfully direct your utmost attention" that finding five hundred recruits "was not going to be without difficulties." Augias offered no fewer than four compelling reasons why recruitment was a bad idea. First, it would be impossible to find volunteers for the railroad in his area. Second, the region was still too politically unstable, and demanding workers was likely to cause further discord. Third, the "sanitation situation" remained dire; "the healthy population has been decimated by sleeping sickness" and they still had no doctor. And finally, the few fit workers in the region were already being used to maintain the roads and to work permanently for local European merchants and industrialists. Recruitment, in short, spelled trouble: "It is certain that this system will bring about unfortunate repercussions from a political point of view."

While Augias littered his communications to his superiors with bureaucratic politesse, his own subordinate, a local administrator in the subdivision of N'Gotto named Georges Pacha, minced far fewer words. Asked to produce 150 recruits, Pacha reminded his superiors that his district was politically volatile and had witnessed recent exoduses of entire communities. Recruiting at the moment, Pacha reported, would be "maladroit and dangerous," and counting on volunteers would be "self-deluding." If forced to recruit anyway, he would "have to decline responsibility for the likely consequences." Augias said he understood Pacha's point entirely. Nonetheless, considering the "primordial interest" of the railroad to the colony, Pacha needed to proceed with "whatever may be the regrettable consequences that inevitably result from it."

The effort to recruit thousands of men in 1925 quickly exposed the administration's utter inability to maintain a works project on the scale of the Congo-Océan. Lieutenant Governor Marchand, who had initially requested five hundred men from the region of Lower Ubangi, soon lowered his sights to 150. And within nine months of his original request, he said he'd be happy with fifty. The administration's faith that it would find "volunteers" masked increasing desperation for human bodies. From Augias's interaction with Pacha, it is clear that lower-level administrators—that is, the ones doing the actual recruiting—understood that they need not find actual *volunteers*. Antonetti, Marchand, and other high-ranking officials showed no interest in knowing the methods that subordinates like Pacha used to procure unwilling recruits.

Details of the recruitment methods would soon be provided by perhaps the most famous writer ever to weigh in on the Congo-Océan, André Gide. Gide, who would win the Nobel Prize for literature in 1947, was already a major literary force in 1925. A renowned novelist, poet, playwright, travel writer, editor, and critic, Gide accepted an invitation from the French government to tour Africa. Long inspired by Joseph Conrad, Gide had for decades wanted to travel there, especially to the Congo, which he imagined to be a state of mind as much as a geographical reality. And as Conrad died the year before Gide's departure, he dedicated his 1927 account of his travels, *Voyage au Congo,* to Conrad. If *Voyage au Congo* was to be Gide's *Heart of Darkness,* then Pacha offered the Frenchman a character study comparable to Conrad's Mr. Kurtz. Like Kurtz, Pacha

André Gide with Marc Allégret in Equatorial Africa, 1926 or 1927.

was to become synonymous with the brutality of the white man left alone in remotest Africa and the inherent corruption of European imperialism.

Despite these similarities, Gide's voyage could hardly have diverged more from Conrad's experiences as a ship's captain in Leopold II's Congo Free State some thirty years earlier. Conrad had come to the Belgian Congo as an aspiring writer in his early thirties; Gide was already in his mid-fifties and a force in French letters. Conrad, who had been the captain of the barge *Roi des Belges,* had worked long and trying hours along the malarial waterways of the Congo; Gide, despite his socialist leanings, traveled with dozens of porters, cooks, tents, furniture, letters of introduction, and the committed attention of the French administration. Conrad had journeyed alone and, indeed, as he had written in his notebooks, "felt very lonely there." Gide's adventures were shared with Marc Allégret, with whom Gide had fallen in love in 1917 when Allégret was sixteen (and Gide was forty-eight). The two Frenchmen expressed what Gide called a "sensual" attraction for "the negro race."

Despite their differences, Gide shared with Conrad the desire to witness colonialism firsthand. And both would be deeply disappointed by the failures of European civilization. For Gide, no soul embodied his disillusionment more clearly than the low-level Pacha. In October 1925 Gide arrived in the young administrator's region of N'Gotto, in the circumscription of Lower Ubangi , on the border between northern Middle Congo and southern Ubangi-Shari, near the main village of Boda. Much of Gide's voyage was made on the backs of the porters who carried him in a *tipoye,* or hammock chair. But he made this particular leg in a car, which caught the eye of the local chief named Semba Ngoto, who would later say that it was his "curiosity" about the car that made him follow Gide to his camp for the night. Semba assumed that Gide, the older of two "Great White Men" in the car, was the governor. Afraid that Pacha's men would punish him, the chief awakened Gide at two o'clock in the morning and recounted his experiences with the regional administrator and his men.

Semba's story made for bleak listening. Shortly before Gide's arrival, Pacha had sent his sergeant, an African named Yemba (later referred to as Niamba in official documents), to punish the inhabitants of the village of Bodembéré for refusing to relocate. For months, Pacha had been trying to move villages closer to the main road that they were expected to maintain as part of their labor tax. The villagers, however, did not want to leave their crops; nor did they, as ethnic Bofi, want to move closer to the road, which was already populated primarily by Baya. To punish the Bofi villages, the sergeant and his armed men undertook a campaign of horrific violence. According to Semba, they randomly rounded up men from a number of local villages, tied them to trees, and shot them. Women and children were burned alive or hacked to death. Allégret took concise notes during the conversation with Semba. Recording the number killed, he jotted: "Men, women and children: 32."

Gide was clearly disturbed by Semba's story of the African guards' brutality, but he insisted that the real culprit was Pacha and, by extension, the local concessionary company he served. Before receiving orders to recruit for the Congo-Océan, Pacha had been helping the Compagnie forestière Sangha-Ubangi, the local concessionary company, to find and discipline rubber collectors. Gide correctly predicted that Pacha and his superiors would pin the blame on the African guards for having "poorly understood or poorly executed his orders." But the guards had been "inspired by the spirit of their master, a somber and sickly man," as Gide put it, "who doesn't hide 'hating the negro' and shows it."

As proof, Gide reported that Pacha had overseen the torture and murder of an African man forced to collect rubber. Villagers who failed to harvest their quota of rubber in Pacha's subdivision were "condemned to circle around the factory, under a leaden sun [carrying] very heavy wooden beams" on their shoulders. If a worker fell during this "ball," then guards would whip or beat him until he got up and continued the macabre waltz. In early September 1925 one such ball got under way in Bambio under Pacha's watchful eye. It was market day, and chiefs and villagers from all around witnessed the humiliation of ten workers. Around eleven a.m. a man named Malongué, from Bagouma, fell to the floor and could not be revived. His body was brought to Pacha, who reportedly said, "I don't give a damn," and made sure the dance continued. Malongué died later that evening.

In Gide's opinion, Pacha's heartlessness in Bambio was but one example of how the administrator mistreated the men, women, and children living in his district. At Pacha's order, workers who brought quantities of rubber that were deemed "insufficient" were thrown into the prison in Boda. The jail was a place of misery and starvation where half of the inmates died; indeed, in one case, twenty villagers entered and only five survived to be released. While men were harvesting rubber, often far from their homes, the task of porterage fell exclusively to women and children. Building and maintaining the main road, which was used about once a month by the district's single car, cost "a large number of human lives." Such examples of abuse caused a "worrisome exodus of the natives" from the region; villages that once counted 135 able-bodied men now had 35, of whom only 15 were able to work.

Gide showed an uncanny understanding of how colonial administrations functioned, especially for someone who had been in country for less than two months. He rightly warned the governor that his African witnesses would be punished as soon as Gide was out of sight. And in fact, before even mailing his letter, Gide heard that Semba had been "thrown in a cell" for speaking out of turn. He also foresaw a response of deep skepticism on the part of the governor. Gide assured him that, in talking with Africans, he had "put himself on guard" against "exaggerations and distortions of the smallest facts." Yes, the accusations stood at odds with what the governor's administrators in the field had reported. But, he pointed out, administrators "present in their reports, by choice, the facts that they think are most likely to please you." Pacha had spread "terror" in his district and needed to be removed immediately.

Gide had no interest in launching a more extensive investigation himself—he had traveling to do. But he clearly knew that his clout, with both the press and the government, would force the administration to investigate for him. Should this not be clear to the governor, Gide spelled it out. "This unfortunate affair," he signed off rather ominously, "risks having the most unfortunate repercussions." The governor needed no goading. An investigation was launched almost immediately, with reports filed before the month was out.

Gide's chance encounter with Semba gave rise to the largest

scandal to hit Equatorial Africa since Pierre Savorgnan de Brazza was dispatched to the region in 1905 to investigate stories of abuses and atrocities. Gide's letter to the governor on the "Pacha Affair" (as the administration itself began to call it) would be reprinted in the French press and, eventually, in his *Voyage au Congo*. The disturbing details uncovered would be debated in the halls of government and on the floor of the Chamber of Deputies. In these discussions, the focus was on the misery wrought by the colony's concessionary companies, especially the Compagnie forestière Sangha-Ubangi. The Congo-Océan was less frequently mentioned, which was not surprising since recruitment had just begun in Pacha's region. But many of the dynamics at work in Lower Ubangi—violence, lawlessness, and minimal French presence and control—revealed the conditions in which the colony's officials lived and worked. And Pacha himself was quick to blame railroad recruitment for the dysfunction of his region.

As horrific as it was, Gide's account proved tame when compared to what official investigations found. Pacha was himself involved in conducting the early investigation—a clear sign of how understaffed the colony was outside Brazzaville. His inquiries revealed violence and intimidation to be more widespread than the future Nobel laureate could have imagined. Though depositions of local witnesses were vague on dates, times, and even locations, they made clear that women and children had recently been burned alive on at least three occasions in the region.

The testimony that villagers gave left little doubt that the slaughter had been planned in advance and carried out with cold determination. In Bogobolo, a forty-five-year-old Bofi woman named Soko told of how she and three children had been forced into a hut by Niamba's men, who set it ablaze. Soko was able to escape from the burning building, only to be held by the guards and beaten. The children left inside were burned alive. Violence was aimed less at killing individuals than in spreading terror. Soko—like many of the victims, it seems—had been chosen at random. A number of the men and women who testified saw the events while hiding in the bush, having fled for their lives.

Fear of the French administration was palpable in the testimonies. A number of witnesses were less than forthcoming with details to their white inquisitors. And Pacha was certainly an awkward choice for an allegedly impartial investigator, deposing men and women who'd witnessed atrocities committed by his own soldiers. In the village of Boubacara, a thirty-year-old Bofi woman named Fotiguela was asked if she had witnessed any women "willfully burned alive in a hut" by Sergeant Niamba. Fotiguela hesitated and seemed uncertain how to answer:

A.  I saw no such thing.
Q.  You have nothing to fear, you must instead tell what you know and above all what you saw.
A.  I saw . . . enclosed in a hut four people, the women Yélé, Konderé, Naco and Uico. This hut was then lit on fire and these women perished like that.

In no instance were the sergeant or his men described as being troubled by the brutality. In Boubacara, one of the guards, a man named Bouendé, lit the hut on fire while Sergeant Niamba simply stood by and watched. "Did the sergeant try to stop him?" Pacha asked a man named Botobolo. "No." In the encampment of Mocogo, a forty-five-year-old Baya man saw guards lock three children in a hut and burn them to death. In that case, a local chief, Kobélé, lit the fire; Niamba and Bouendé did nothing. "Neither the guard nor the sergeant opposed this crime?" Pacha asked. "No," replied Babodji, "they all agreed with it."

Pacha was soon removed from the process, and an inspector

named Jean Marchessou took on the case. Marchessou defined the old colonial hand: he had been working in Equatorial Africa since 1905, had received multiple commendations and awards, and was considered intelligent, strong, and tireless despite spending years in insalubrious posts. Like Pacha's, his investigation focused on the misdeeds of African auxiliaries. Marchessou implicated Niamba in further acts of brutality, including killing babies. In one case, he tore an infant, Ngounou, from his mother's arms and slammed him "brutally" to the ground. He then ordered one of his guards, Bengué, to dispose of the baby's body in the bush. (Niamba later denied this account, nonetheless belying a certain indifference to the cause of death: "I put [Ngounou] on the ground and, if he died, it's Bengué who killed him.") On another occasion, the sergeant oversaw the murder of three small children, all killed with spears. He also ordered the murder of the wife of a man whom he claimed had injured one of his men. But clearly he needed no real justification to kill: in one village he whipped an unarmed man named Bili, then shot him dead, as one witness noted, "for no reason."

Marchessou's findings were wide-sweeping and implicated Pacha, not specifically in the ruthless actions of Niamba and his men but in the general brutality with which the French state engaged with the local population of the entire region. What represented acceptable or unacceptable behavior in such parts of the empire was a matter of judgment. The code of colonial laws that ruled indigenous populations, known as the Indigénat, lent tremendous leeway to French officials on the ground. Rather than dictate relations between administrators and subject populations, colonial law often followed whatever practices officials deemed necessary. Even according to a system that gave officials the benefit of the doubt, Pacha's behavior seemed excessive.

Marchessou found that the "ball" that had killed the rubber harvester Malongué was not a unique event; torture and imprisonment

were regular punishments for "insufficient" work. The prison that
Pacha administered had a "regrettable reputation," where inmates
were beaten and denied food and medical treatment. At least one
inmate had starved to death, while a second died at the hands of
Niamba. All the inmates bore the scars of the whip. "The responsibility for this affair is entirely Mr. Pacha's," Marchessou determined, "at the very least for the inhuman and illegal punishment
he inflicted." He stopped short of blaming all the problems in the
region on Pacha. The area, for example, had lost one-third of its
population since the previous year. Marchessou did not blame Pacha
for the depopulation, though, noting that "exoduses" had happened
in "different epochs" for purely "local reasons."

Another of Pacha's superiors, a Mr. Jacoulet, was even less for-
giving. Admitting that Pacha was "young and insufficiently experienced" with the "extremely delicate" task of administering a difficult subdivision, Jacoulet nonetheless claimed—echoing Gide—that
Pacha's "dominant trait" was "hatred" of the indigenous population.
As a result, Pacha rarely engaged with local people and left his soldiers to act as they pleased. He allowed internecine quarrels to rage
among villagers. With murder and pillaging commonplace, "human
life counted for nothing."

⸻

THE INVESTIGATION INTO PACHA'S brutality produced a
number of reports that shed light on French rule in places like
Lower Ubangi in the mid-1920s. Pacha, true to form of so many
administrators, was a young and inexperienced man in a region
that he himself described as "paradoxically vast." Born in Nîmes in
the South of France, Georges Pacha was a decorated veteran of the
First World War who joined the colonial service in 1920 and went
to Equatorial Africa the following year. His reasons for choosing a
life in the colonies were never fully explained, but the birth of an

illegitimate son in his hometown might have offered some motivation. Paternity suits aside, Pacha rose steadily in the bureaucracy
and, after four years of low-level positions, found himself in charge
of a region in northern Middle Congo that Frenchmen associated
with warring chiefs, violent instability, and cannibalism. Indeed,
as late as the spring of 1925, Pacha's region remained, in the words
of one official, "in open anarchy" and completely "refractory" to
French power.

The inexperienced administrator was provided with minimal
direction from his superiors. The advice he did receive was regularly veiled in euphemism that thinly covered the violence he was
clearly expected to employ. The need for French administrators like
Pacha to be "tough" (*dur*) was a common theme throughout Equatorial Africa. Pacha himself made toughness the cornerstone of his
legal defense, claiming that his superiors had told him to put an
end to rebellion in his region "by all means" and to show himself "always very tough with the natives of N'Gotto." He followed
orders. His personnel reviews, which from this period were very
positive, described Pacha as "firm but fair"; he received one such
review just a week after the Bambio "ball" where Malongué died.

Pacha's tactics against rebellious chiefs in his region illustrate
what administrators understood as being "tough." In May 1925
Pacha set out to change the balance of political power in his subdivision and bring warring groups under his authority. About fifteen
kilometers north of the regional capital of Boda, he and a small
group of soldiers encountered, nailed to a post, the tattered skin of
a guard who'd been killed the previous year by rebels. Pacha had
received a message that he should come claim the human hide "if he
had the courage for it." Allegedly outraged by this "savagery," Pacha
attacked the rebels with guns blazing. What followed was the "systematic destruction" of all the settlements in the bush. Pacha then
undertook to relocate the population to new villages (and likely to

destroy their existing homes) located closer to the road to facilitate surveillance and labor recruitment.

Over the following two months, Pacha faced continued resistance from a number of Baya chiefs. The most serious attacks were led by a local leader named Bapélé, who killed his enemies, burned a village, and destroyed crops in an act of defiance against the "commandant" who had challenged him. The Baya allegedly ate the flesh of those they killed. Whether claims of cannibalism were accurate doesn't alter the reality that they worried French administrators in the region. With the support of his soldiers and a group of local partisans, Pacha led a counteroffensive, killing fifty Baya, including four chiefs. More were taken prisoner, including a chief who was later killed in prison by Sergeant Niamba. The Baya did not relent, continuing their attacks and massacring around a hundred villagers in various skirmishes. Pacha countered as best he could; as his superior put it with another flurry of euphemism, "justice was severely exercised."

Pacha's methods were clearly shaped by specific local political conditions. But in reflecting on his actions, a number of officials pointed out that they were commonplace. Mr. Augias, in an early attempt to defend Pacha, spoke in surprisingly frank, if not always eloquent, terms. "I am not a torturer, far from it, I love the native and I am loved by him," he began, with a disclaimer that might well have worried his more expectant readers. "And yet I reckon that it's necessary to act with the greatest firmness when the circumstances call for it and these circumstances are unfortunately not rare in the circumscription of Lower Ubangi." With Gide in mind, the administrator admitted that such treatment shocked "travelers coming from the metropole" who were "upset a little too quickly." Outsiders were ignorant of "the considerable difficulties encountered by bureaucrats charged with establishing order in immense territories where the means put at their disposal are generally very insufficient."

Augias's observations revealed a truth only too well known to administrators in Equatorial Africa: the colony was a massive expanse of lands, climates, peoples, cultures, and languages that, in the 1920s, was only tentatively administered by the French state. Orders emanating from Brazzaville went out to a colonial bureaucracy that made sense on the map neatly marked with administrative regions, subdivisions, and towns. The French penchant for administrators, combined with a reluctance to give local chiefs real power, meant that there were about three times as many white functionaries in French possessions as in British colonies. But this did not translate either into more effective leadership or into greater interaction between officials and Africans—indeed, quite the opposite. Equatorial Africa was such a large colony that even hundreds of administrators provided only the thinnest presence on the ground.

Making matters worse, stark differences of race, language, and culture led many white administrators to barricade themselves from their local communities. The idealized colonial image of a white administrator engaging with inquisitive African subjects was far from the norm in an increasingly bureaucratic empire. Certainly individual officials had relations with local men and women that ran the complicated gamut from servant to lover. But as a matter of professional practice, proximity was avoided. Even colonial inspectors whose job was to tour and interview communities about social, political, and economic conditions and report to the ministry could show little interest in meeting with Africans.

A 1905 postcard of a visit made by "Administrator E." in the Congo captured the nature of exchanges between officials and Africans. In the photo, the gaunt Monsieur E.—decked out in white linen suit and pith helmet—glowers in the direction of the camera. Dispersed about him are an African child and three adults, as well as another Frenchman, most likely a local resident. On this day, there are no jovial meetings between colonial official and subject to be

A French administrator in French Congo, postcard, c. 1905.

found, no curiosity, no exchanges of ideas or anything else. Instead the image reveals a world of fatigue and inertia, of scattered souls gazing in different directions, transfixed in psychological isolation. The poor child seated behind the inspector's back, face propped in hand, looks bored senseless. If the facial expressions aren't revealing enough, the photo highlights the more casual divisions of empire as well: the two European men sit in chairs, while the Africans are left to stand or sit in the dirt. The two white men sit in the foreground; the two women in the scene remain hidden in the distant shadows.

By the mid-1920s, the administrator's life in Lower Ubangi had certainly not become much more appealing than that of "Administrator E." For over a decade, the region had experienced "unceasing difficulties," which Augias claimed, in rather awkward terms, stemmed from "the ceaseless growing shortage of administrative personnel." His numbers spoke more eloquently than his words: before the First World War, the region had had twenty-six French administrators, but that number had dropped to *five* by the time Augias arrived in 1925. Two of these five went on leave soon after Augias's arrival, despite the fact that Antonetti had deemed seven administrators to be the absolute minimum needed to rule. The three remaining officials were in poor shape: Pacha suffered from fatigue that rendered him ineffective; a second administrator was "seriously fatigued" and had been hospitalized in Brazzaville; and the last man was a recent arrival who despite his "good will" could not cover for all the absent administrators.

Augias, then, effectively took over all administrative responsibilities of a vast region that spread over an area of roughly 8,500 square miles (22,000 square kilometers), about the size of the U.S. state of New Jersey. Unlike the Garden State, this corner of Equatorial Africa had minimal infrastructure, forcing administrators to follow winding waterways or footpaths through dense undergrowth when traveling from village to village. Augias tried to spend twenty days per month touring to get a sense of what was going on with the local population, but the effort was always "very painful, in generally unhealthy, swampy regions." The ten days per month he spent at his own post were no better. The endless official duties he had to tend to were a "veritable hell." His constant exertion to keep on top of his responsibilities left him "winded" and his health "shaken," causing him certain anxiety.

Should Augias have exaggerated his circumstances, Pacha's account of life in Lower Ubangi wasn't much different. Before being posted to the subdivision of N'Gotto, he had spent eighteen months in Impfondo—a "painful period" he claimed, that "wore him out." Once in Boda, Pacha endured four months of "uninterrupted physical action" before suffering sunstroke, the effects of which he dealt with for months. At this time, his young daughter in France died, which did little to help him recover. While he did not mention it explicitly, the sheer loneliness of being the only Frenchmen in the area—his superior, Augias, was a circuitous 500-kilometer trek to the south—could not help but take a psychological toll. Regardless, expectations remained. Pacha was ordered to subdue local groups that burned villages and murdered their neighbors, to collect taxes from the inhabitants of N'Gotto, to increase production of manioc and rubber, and to recruit labor for the local concessionary company and the Congo-Océan.

Considering the extraordinary burdens placed on the twenty-six-year-old, it is easy to understand why Pacha did not take kindly to the criticisms of André Gide, who showed up one day, moved into his administrative building, started ordering his guards around, called meetings with local chiefs, borrowed his car, and even instructed him one morning to make some omelets for breakfast.

When it came time to defend himself against accusations of abuse, Pacha reiterated that violence was a necessary part of administering subdivisions like his own. Upon his arrival in Boda, he said he encountered "an exceptional situation implying exceptional measures" and was ordered to bring a difficult region under control. Unapologetically, he admitted that "strictly regular measures" would not have brought "satisfaction"—indeed, employing such measures would have been "a veritable anachronism." Experience showed that "extreme timidity toward rebels" garnered only "limited results." Did the trying conditions and extreme instability push Pacha to use "brutal and irregular means"? the young administrator asked rhetorically. "I can only regret so; but"—and the *but* was crucial—"I was alone where there used to be two [administrators]. It would be just to remember that."

The court and colony did remember it. There were three judicial proceedings in late 1926 to hear criminal charges in the affair: Pacha faced charges in two hearings before the *tribunal correctionnel;* the African soldiers were tried before the *tribunal indigène.* Pacha was found guilty of negligent homicide (*homicide par imprudence*) in the case of Malongué at the Bambio "ball" and punished with a one-month suspended sentence and a fee of five hundred francs. For general abuse of authority and misuse of violence, he was given a further one-year suspended sentence. In other words, though he was found guilty and fined, he never saw the inside of a prison cell. The magistrate who oversaw the proceedings noted that Pacha's main argument of defense—that he was repeatedly told to be "tough" with the local population—was not entirely convincing. Violence could be justifiably used only with "refractory" ethnic groups, not with harvesters or prisoners. But he was moved by Pacha's youth and inexperience, as well as the turbulent nature of his region, as mitigating factors.

Pacha was recalled from Boda to Brazzaville, reassigned, and rehabilitated. A few years later, a new colonial minister, appalled to learn that Pacha had never faced real punishment, pressured the governor-general, but Antonetti defended his subordinate vigorously. He blamed the entire episode on Pacha's superior, Augias, who was eventually (and perhaps for Antonetti, conveniently) killed in Lower Ubangi. In Antonetti's assessment, Augias had not effectively advised his young and inexperienced subordinate. The governor-general recast the entire affair as a minor breakdown in administrative practice, not a case of colonial atrocity and murder. Pacha went on to have a long and successful administrative career.

Pacha's men received no such leniency: Sergeant Niamba and the guards were implicated in the deaths of twenty-three Africans. Niamba, three guards, and one local chief received life sentences for murder. Eleven other soldiers received sentences of five to fifteen years in prison. Pacha had claimed his men used violence without his knowledge. Niamba had insisted Pacha ordered him to carry out the attacks on the local population. Pacha's word was validated; Niamba's was not. Such was the color of justice in Equatorial Africa.

THE CONNECTION BETWEEN THE "Pacha Affair" and recruitment for the Congo-Océan was disputed at the time, most vigorously by Antonetti. The governor-general acted as though the brutality uncovered had nothing to do with construction of his railroad. But Marchessou's investigation found that the tour that resulted in the murder of the women and children by Sergeant Niamba's troops—the tour uncovered by André Gide—also netted three to four hundred potential recruits for the Congo-Océan. Back in Boda, these people were presented to Pacha, who chose a hundred of them for the railroad. They remained in Boda for about a month—in a prison, as was a common practice with recruits for the railroad—and were then sent by boat and on foot to the construction site in early December 1925. These were the "volunteers" that Pacha found to fulfill the quota demanded by his superiors in Brazzaville.

The Congo-Océan recruits were taken with the use of "physical constraints"—again a euphemism for force. Pacha used the opportunity "to clear the country of undesirable elements," selectively taking men he deemed to be troublemakers in the region. He executed the order in the same way many French officials motivated rubber harvesters and other workers for the concessionary companies: he sent his sergeant and three guards, all armed, to round up men for the railway, and to lead the women, children, and elderly to a camp designated for their resettlement. He knew that there would be violence, noting that Niamba and his men felt "exasperation and hate" for the "savagery" of the "primitive" races. Administrators like Pacha relied on violence to rule. Recruitment for the Congo-Océan, especially in regions where the French hold on power was tenuous, would be administered in the same way.

Pacha, in fact, admitted to the link between murder and recruitment, saying it was impossible to recruit for the Congo-Océan without resorting to coercion. He sought the legal help of a Mr. Amouroux, the president of the Brazzaville chapter of the League of the Rights of Man (Ligue des droits de l'homme), a major French civil rights organization. In turn, Amouroux reported to league headquarters in Paris that recruits for the railroad were clearly, as he wrote it, "NON-VOLUNTARY needless to say." "In this recruitment of workers," he continued, "which strangely resembled an organized manhunt, more than thirty men or women had been burned alive, or killed, the fatal results of received orders." Shifting the focus away from the administrator, the league insisted that the real issue was the method employed for the recruitment of colonial labor, "especially for the railroad construction sites." For the league, then, the Pacha Affair could not be understood without reference to the Congo-Océan.

For Antonetti, the links between the violence of the Pacha Affair and the recruitment of workers for the Congo-Océan represented an early challenge to his effort to make the railroad his main priority. His responses to both the league and the Colonial Ministry betrayed his relative inexperience with criticism. Many of the elements that would come to shape his repertoire of responses were evident in the months after events in Lower Ubangi came to light. Front and center was denial, plain and simple. Antonetti insisted there was "manifestly no connection" between "the recruitment of workers for the railroad" and the "atrocities" committed by Pacha's men. This was a claim he made repeatedly; but he knew it to be false from Marchessou's report, which found that Niamba's atrocities had been committed on the same tour that rendered a hundred recruits for the railroad.

The violence unveiled in Pacha's region was strikingly reminiscent of the accusations made against concessionary companies a quarter-century earlier. Pacha's brutal mistreatment of men, women, and children was not unlike the atrocities documented in the Brazza commission's report. The investigation into Pacha showed that, in some parts of the colony, methods of labor conscription had changed very little. Such a connection between the Congo-Océan and the concessionary past did not bode well for a governor-general who had promised a bold future and a stark break from the mistakes of the past.

If the colony's stated hopes and goals had changed, the means of attaining them had not. As Albert Londres pointed out a couple of years after Gide's visit, the African's experience of colonialism was far from one of paternalist goodwill. And the aspiration of achieving a richer, more peaceful colony together—French and African—bore little resemblance to conditions on the ground. "The administration is the mosquito of the negro," Londres wrote. "At every moment of his life, it stings him, troubles his rest." When the administration demanded workers for any work project, be it collecting rubber or repairing roads, Londres observed, "it's captives who do it."

# CHAPTER 4

# MANHUNT

For man also knoweth not his time: as the fishes that are
taken in an evil net, and as the birds that are caught in the
snare; so are the sons of men snared in an evil time, when it
falleth suddenly upon them.

ECCLESIASTES 9:12

TENS OF THOUSANDS OF the men and women who worked on the railroad followed a similar path to get there. In September 1927, to draw
on one man's experience as an example, a thirty-year-old named
Malemale was taken from his village of Balla-Kité in Ubangi-Shari
to a medical inspection camp near the capital of Bangui. Almost
nothing is known about Malemale's life, except that he was married to a woman named Vlapedoum and was the son of Batakoudou
and Gbaké. At the medical camp, a French doctor named Kerjean
weighed, measured, and examined Malemale. At 1.8 meters and 57
kilos (about 5 feet 10 inches and 125 pounds), Malemale was already
severely underweight. Dr. Kerjean had inspected many underweight
recruits but was expected to fill high quotas of laborers. He deemed
Malemale "capable" for work on the Congo-Océan.

In the medical reports documenting Malemale's short life as a recruit, the next five months are blank. During that time, he was transferred from the medical camp to Brazzaville, more than a thousand kilometers to the south. There Malemale underwent another exam; again, a doctor found him physically capable of work. Two weeks later he arrived in Mavouadi, near the railroad construction site, probably having walked more than two hundred kilometers from Brazzaville. There a doctor found him unfit to work due to "insufficiently muscular legs." His body was wasting away, probably due to malnutrition or disease. Malemale was admitted to the small bush hospital, where he died the next day. He had spent months traveling some two thousand kilometers across the center of Africa never to work a day on the Congo-Océan.

Between 1925 and 1932, when recruitment ended, some 42,000 men and women would survive transportation from Ubangi-Shari to Brazzaville; at least 5,000 would make it from Chad to the railroad. Antonetti's decision to recruit in the north not only opened up a much larger labor pool, it also allowed recruitment among prized ethnic groups. The French believed the inhabitants of the north of the colony, whom they imagined to be bigger and stronger than the inhabitants of the southern forests, to be particularly well suited to the task.

Recruitment aimed to find strong, healthy men who could perform hours of strenuous manual labor daily, and who would agree to be away from home for a year or more. A photograph from 1926, taken during the tour of a Captain Paul Lacoste, an official recruiter who prided himself on the quality of men he could find, showed the ideal type of recruits. It is unclear what purpose the photograph served, but it presented the men—tall, muscular, dressed, and possessing an air of military discipline—in a positive light. Men were clearly the most sought after, but between 5 and 10 percent of the recruits were women, brought in less to perform heavy manual labor

Workers from Chad or Ubangi-Shari, with a French administrator, c. 1926.

than to provide support cooking, cleaning, and tending to camps. Many women were wives who joined recruited husbands.

The traces left of Malemale's journey are but a general template of what tens of thousands of men and women lived through. By pairing it with other accounts and reports, providing both general overviews and highly specific incidents, a more detailed rendering of the odyssey comes into view. The motivations and experiences of so many people confound easy generalizations. But historical sources tell of a range of conditions, episodes, and treatments that many, if not all, men and women encountered. Some certainties become clear. Recruitment was not a peaceful or easy process. In terms of intimidation and violence, the Pacha Affair was clearly not an isolated case; rather, brutality continued to plague recruitment methods until 1932, when the hunt for men finally came to an end.

Interaction between potential workers and the administration began with chiefs, the intermediaries who were supposed to deliver men without conflict. Antonetti himself, in a beautifully printed circular sent to his administrators in the field (and likely produced for public consumption as well), offered direction on how to entice men to join the Congo-Océan cause. Only the railroad could bring would-be recruits "cheap merchandise." He encouraged spreading the word that workers were paid and fed as well as possible: "If they are sick, they will be cared for." Everything that could be done to make their work "less hard" would be done. Of course, Antonetti added that should this reassurance not calm their concerns, then recruiters could always resort to reminding them of the law: *"You owe us your help with our works projects."* Should his italics not be blunt enough, Antonetti made his point with a footnote describing the 1919 decree that legitimized "obligatory" labor—the politer term for *forced labor* that many colonial apologists preferred.

Certainly, many men and women volunteered for a variety of reasons, finding the promised pay attractive or aching to get away from home. Some likely agreed to recruitment in order to travel with friends or family. In his 1987 novel *The Fire of Origins*, the Congolese author Emmanuel Dongala tells the story of Mankunku, a young man coming of age at the advent of French colonial rule. Mankunku agrees to work on the railroad because he has grown wary of hunting for rubber plants for the concessionary companies. With the hope of earning enough to pay his tired parents' taxes, he agrees to take the job. Some men and women no doubt made similar rational decisions. For individuals who sought independence from kinship obligations, the railroad offered a financially tempting change, be it permanent or temporary.

Rational decision making, however, could be manipulated, and many recruits had little idea of what they were signing up for. Aside from the reality that few recruits could imagine what life on a distant construction site would be like, active deception by recruiters was commonplace. Indeed, lies about what lay ahead for recruits were widespread enough for Antonetti to admonish his subalterns against tricking workers into believing they were being recruited for military or administrative work. "You must tell them the truth," he ordered. There is, however, little evidence of how effective his message was.

If recruitment provided an opportunity for some, it struck many others as a threat. Men and women regularly fled recruiters—to the bush, onto plantations, and even into neighboring colonies—in order to avoid going to the Congo-Océan. It wasn't the work itself that they feared; it was the railroad. One administrator found that if he needed workers to maintain a local road, he had no trouble finding people; by contrast, recruitment for the railroad sent men running. In certain regions of Lower Congo, inhabitants lived in encampments specially designed to allow them to escape recruitment at a moment's notice. Only the sickliest were unable to flee at the first sight of recruiters, a fact that helped explain why so many unhealthy men seemed to be recruited. As one inspector put it, "The prospect of the construction site isn't appealing to anyone." The one possible exception to this rule was the population around the railroad who were familiar with the region and, in some cases, could live at home.

Farther afield, recruitment was often avoided at all costs. In southern Chad, life came to a standstill when recruiters came around. "All the fit men are hidden in the bush," one administrator wrote, "the villages and fields are deserted." Rumors of high mortality rates and difficult working conditions, especially in the Mayombe, spread across the colony with remarkable speed. One French traveler said that in central Ubangi-Shari, the mere muttering of the word *Mayombe* caused the "natives to hide their heads in their arms shouting."

Practical concerns about health, welfare, and distance from family and loved ones found expression in local belief systems. A common myth among inland societies was that the ocean harbored death. Sirens, called *mamiwata* in one manifestation of the legend, who possessed the lower body of a fish and the upper body of a white woman with long hair, were believed to draw men to the water. The very sight of the ocean—possibly a metaphor for the historical reality that saw thousands taken in slave caravans to the coast of the Kingdom of Loango—was thought to be deadly. Recruits from the interior of the colony believed—rightly, in some cases—that working on the Congo-Océan would force them to the haunted sea.

If mistrust, rumor, and myth weren't reason enough to fear the Congo-Océan, the brutal methods employed by recruiters spread terror across the countryside. Despite orders to the contrary, recruitment was regularly carried out in the presence of armed soldiers (*gardes regionaux* or *miliciens*). As early as 1923, officials warned that the presence of soldiers would interfere with the proper use of chiefs as intermediaries. But for years such admonitions failed to dictate practices. Methods of intimidation dating back to the heyday of the concessionary system were still used. Armed soldiers rounded up women and children as hostages until the men came forward and volunteered. When soldiers took women, it often meant sexual abuse. As one man explained to a French traveler who was curious where the young women of the village had gone, "the *miliciens* [soldiers] took them to their huts. At night, they dance for them. Then they 'sleep with' [them]."

Soldiers also intimidated uncooperative chiefs, who could be whipped, imprisoned, or worse, escalating tensions. Some chiefs resisted by encouraging their villagers to flee. Others stood up to administrators who asked for more and more men by simply refusing to cooperate or by insisting their regions were "spent." In at least one case, an unnamed chief from Boko-Sangho chose to face death rather than continue to send his kinfolk to work. The French administration replaced recalcitrant chiefs, in an effort to reorganize the political landscape to ease recruitment. But French-appointed chiefs enjoyed little or no respect from the communities they ruled, causing more social strife rather than less. Appointed chiefs sometimes used violence against their own people to find workers.

Whether locally respected or not, chiefs were unreliable intermediaries for officials seeking quality labor. Chiefs regularly filled recruitment quotas with men of dubious quality. To improve their own political situation, they offered political adversaries as recruits. Or they assigned men who were wholly unfit for labor—too weak, too young or old, too sick, or known to be troublemakers. Other kinds of corruption could also be at work, including would-be recruits paying men to go in their stead, or ordering slaves or debtors to take their places.

In some instances, villagers turned on their chiefs and recruiting agents, resulting in serious injuries. An administrator named Montchamp, posted in the region of Baguirmi, reported on a desperate situation in southern Chad and Middle Shari in 1928 and 1929. In Djimber, a group of people, ranging from children to the elderly, all of whom were afraid of recruitment, fled to caves in a mountain high above their village. When soldiers tried to apprehend them, the men rained stones down upon the troops, cracking open a head, breaking a leg, and crushing the feet of three different soldiers. In the village of Nyio, a group of villagers attacked a recruiter in order to free some ninety recruits he held captive; one guard was injured with a knife and one aggressor was shot and killed. Chiefs sometimes fought back: at Mossala, a chief killed six and injured more than a dozen in an effort to escape from an attack. When another chief was attacked, Montchamp reported bluntly, in terms as vague as they were ineloquent: "There was death of men." On another occasion, a corporal was killed; on yet another, a Sara recruit committed suicide to avoid being taken by *la machine*.

Such conditions made Montchamp circumspect about recruit-

ment, if not entirely intimidated by the prospect. He drew the ire
of his superiors by refusing to recruit among the Muslim population. Doing so, he argued, would throw the region into "open
resistance"; such an uprising could be quelled, he continued, but not
without causing a "severe exodus towards the neighboring English
colonies." He did have some hope for the future. Some recruits from
the region had recently returned and had told their families they
were well treated on the Congo-Océan. But in making the point,
he betrayed the methods that were most often used to recruit: "I
continue to think that it would be good to give up recruiting by
force as we are obliged to do."

The combination of armed recruiters, unreliable chiefs, and
anxious villagers took a toll on the countryside. Whole areas were
depopulated as communities took flight before recruitment could
take place. An administrator touring northern Middle Congo in
late 1927 reported that, following recruitment in one town, the
neighboring villages cleared out, fearing they were next on the
list. "Since Makoua," he wrote, "the villages we passed through
(Dounga, Makounbé, Assagnan, Laboi) are deserts." At best, the
official encountered "chiefs and some women." Only one village
chief presented workers to the recruiters; armed guards gathered the
rest of the volunteers "by employing force." Another administrator
reported in 1928 that when recruitment took place in his region,
near the Cameroonian border, emigration of the population was certain to follow. The same year there were reports of an exodus into
British-controlled Darfur and Nigeria to avoid troops looking for
men.

Accounts of recruitment made their way to French newspapers. An article in the left-leaning *Le Populaire* claimed that stories
of improper recruitment abounded in the colony. An "old colonist"
recounted a story of an administrator who attracted local men to a
celebration of July 14—Bastille Day—by offering copious food and

palm wine. After hours of drinking, men passed out "dead drunk," only to awaken tied by their hands and feet aboard a boat heading downstream to the railroad. This, he noted, was how an administrator chose to commemorate "the French national celebration, the celebration of the taking of the Bastille, the celebration of liberty over despotism, the celebration of humanity after all."

Stories of palm wine and trickery, however, were less common than simple accounts of brutality. Even the staunchly pro-empire *Les Annales coloniales* reported the violence that accompanied recruitment, calling out Antonetti for being either ignorant or deceptive. The article was supplemented with passages from letters sent by a number of Frenchmen in Equatorial Africa who were growing concerned by the continued mistreatment of recruits. The primary fear was that "the brutalities, the exactions" of recruitment in Ubangi-Shari would lead to widespread rebellions against colonial rule and would require military intervention. Reprisals risked destroying the population already tested by years of violence, labor recruitment, and disease. "The intensive and rash recruitment of workers for the worksites of the Brazzaville-Océan takes from villages everyone young and healthy," one letter reported. "Births fall to zero; who moreover, in these conditions, will make babies?"

The reports continued for years after the initial shock of the intensified recruitment, with stories of violence flowing through the colonial bureaucracy. Certain refrains were common: that recruitment was achieved only with extreme difficulty, that villages were left empty, that men were "intensely scared of being sent to the worksites." It is impossible to measure the exact extent to which brutality accompanied recruitment. Certainly, men volunteered for myriad reasons. There were chiefs that handed over the quota of manpower without issue. But the prevalence of accounts from disparate locales that spoke of soldiers, abandoned villages, and desperation suggests that recruitment bore many of the hallmarks of the concessionary system. If Antonetti's workforce was truly free and voluntary, why did entire villages abandon their homes and crops— their livelihoods—to hide in the countryside?

———

ONCE LABORERS WERE FOUND, their interaction with *la machine* had only begun. As Malemale's experiences suggest, it often took months for men to make the voyage to the worksite. Under the best of conditions, men and women endured long marches, from post to post, through changing ecosystems and unpredictable weather, covering hundreds of kilometers and lasting days, weeks, and even months. It was a voyage of privation, abuse, and emotional strain.

Pierre Contet, a writer about whom little is known other than that he spent years living in the colony and spoke native languages, painted a vivid image of recruitment. In 1932, in the waning months of the railroad's manhunt in the north of the colony, Contet encountered a column of African men that led him to believe that recruitment methods had changed little since the days of Pacha. One late afternoon in a village in Ubangi-Shari, Contet saw a number of men approach him. As they neared, he noticed what one sees "in tracked animals about to be taken by a pitiless hunter." When the dust settled around them, Contet could see the men's bloodied bare feet and the cord tied around each man's neck, linking one to the others in columns. They were "haggard, thin, men with imposing, bony frames"; Contet took them for criminals.

"What did these men do," Contet asked a villager standing nearby.

"Them? Nothing!," the villager replied. "Those are railroad workers."

Contet spoke with one recruit, a Sara—the ethnic group many French believed to be strong and well suited to physical labor— named Bamba. Contet asked Bamba where he was going.

"I don't know," he answered.

"How do you not know?" Contet asked. "You left your village without knowing the purpose of your trip?"

"No! I was told I had to work on the white man's 'Kontou-Kontou' (railroad) down in the south. One night, about three weeks ago, the local administrator's soldiers overran all the huts of my village. We were forced out by whips of the *chicotte*. . . . Then we were thrown in prison." The recruits were later led to the local administrator, who examined them and chose ten among the prisoners to be sent to the south. Two days after that, Bamba's group was moved to the prefecture and mixed with other men, some of whom were criminals. They were later tied together by the necks and led on the weeks-long trip to the south.

Contet's account provides the sort of detail and narrative coherence that are common to journalism; but it corresponded to descriptions of transportation found in administrative reports. Bamba's recruitment would have been familiar to anyone who'd paid attention to recruitment methods throughout the period. Once workers were found—whether they had been volunteered into service by their village chiefs or forced by whip or gun—they faced a long and miserable trip to the worksite, sometimes more than a thousand miles to the south, in a world completely unknown to them.

The first stop was a gathering point, often in a large regional town, which could be a days- or weeks-long trek across deserts, forests, or savannahs from their point of recruitment. Once there, they could wait for weeks more, living in prisons or makeshift camps. It was in these posts that workers were first truly processed by *la machine*—the great system of acquiring, using, and regurgitating human beings that was the Congo-Océan.

Bamba followed a path similar to that of Malemale and thousands of other men and women. Once they reached a regional post, they were inspected by doctors. With quotas needing to be filled, doctors were encouraged to accept all the recruits they could. They did, however, determine some potential workers to be *inapte* (unfit) or *malingre* (sickly). Unfit workers were supposed to be dismissed immediately. Men who were considered *malingre* would often be sent on, with the expectation they might perform less arduous tasks on the railroad line. Men deemed healthy were bound for hard labor.

The relatively lax medical standards, however, meant that many men were sent on against all good judgment. One study completed in 1925 suggested that about three-quarters of all recruits should have been turned away, if the most "thorough" exams had been given. More than two years later, inspectors still complained that too many unhealthy men were being sent south on a voyage they were ill prepared to endure. The consequences were dire. From October 1927 to February 1928, between 13 and 52 percent of recruits from Bangui were either seriously ill or dead by the time they arrived in Brazzaville. In February a convoy of 327 recruits arrived from Bangui with only 47 men able to work.

Medical records spoke for themselves, revealing that exams were often less than clinical. To cite just one instance, a young man named Gazaio apparently shrank eight centimeters (more than three inches) in height and gained four kilos (nearly nine pounds) in weight between his first and second medical exams, if his file is to be believed. More curious still, through his first two medical inspections, Gazaio was said to be in fine shape. But on his third exam, after weeks of waiting and travel, he was turned away for being too young.

In addition to being medically assessed, recruits were bureaucratically processed. Recruitment meant not only the loss of home and the social identity provided by family, village, and culture; it also stripped them of individual identity. If recruiters stole their dignity, the bureaucratic machinery took their names. Once men were deemed fit for service, they were assigned an administrative number (*numéro de matricule*) that identified them in official correspondence. A worker, for example, would then be referred to as "No. Mle. 8846." The language of colonial bureaucracy denied individual humanity. Even in general reports, instead of speaking of boys or men, officials used generic terms like *recruits, workers, units (unités),* and *workforce (effectifs).*

Once they were inspected and numbered, sufficiently healthy men and women were meant to be issued a sleeping mat and clothing for the voyage, but few it seems were. While some accounts discussed the use of mats, very few suggest that clothing was issued. Indeed, sufficient garments were not available even on the worksites until the late 1920s. With minimal clothing, no blankets, and often no personal possessions, they began the procession to the south. Marched with armed guards, recruits' lives were no longer their own.

A recruit's sense of perdition was made complete when he was chained or tied to other recruits. Few images evoked comparisons to Atlantic slavery more than the shackling of men, especially binding them together by the neck in a coffle. And yet many recruits were so bound. The practice was prevalent and persistent enough for the governor-general to issue multiple orders prohibiting it. The binding of workers had technically been outlawed since the 1890s. But in 1926, more than a year into the expansion of recruitment to the north, Antonetti reminded his officers of the rule. Then, in 1928, the practice was still common enough to make the governor-general feel the need to send out a circular making chaining "formally prohibited." He then in mid-1929 described himself as "painfully astonished" to read a casual reference to the chaining of recruits. By that time, his administration had begun to insist that recruitment no longer required force or intimidation. Of course, considering Contet's later account, the governor-general's reiterated prohibition in 1929 seems to have been as effective at stopping the practice as had been the original regulation aimed at concessionary recruiters.

The reference that caught Antonetti's eye in 1929 appeared in an investigative report in which the tying of men together was perhaps one of the less alarming details. In April 1929, a convoy of recruits in Middle Congo came to a stop when a number of men tried to escape. With the sergeant, a man named Meto, in pursuit, a soldier named Kondo checked to make sure the cords tied around the recruits' necks were secure. When Kondo came to check the cord of a recruit named Ondema, he found it unlocked. "You have undone your rope," Kondo reportedly said, "because you wanted to escape like those from your village just did." Ondema said he had no intention of escaping and made not the slightest menacing gesture. But Kondo shot him point-blank. The bullet pierced Ondema's heart, and he fell dead, pulling down two other men who were bound to him. Kondo cut Ondema's body loose and threw his body to the side of the road. He then threatened to kill the next person who tried to escape.

The response of various administrators to Kondo's murder of Ondema revealed many things about the transport of recruits. Antonetti showed far more concern for what the murder would look like to the outside world than for the violence committed. He berated his subalterns, ranting that he found it "incomprehensible" that administrators would allow this sort of thing to happen considering the colony's "unfortunate reputation" for "alleged brutality." Tellingly, though, Antonetti showed no concern that armed soldiers were being used, despite policies to the contrary dating back to at least 1923. Since Kondo was armed with the weapon that killed Ondema, the convoy of troops must have looked essentially like the one Pierre Contet would encounter in a different region a few years later.

Equally revealing, the lower-ranking officials who commented on the Kondo case in their own correspondence showed no concern that the convoy of men had been tied together in a coffle, suggesting it was not out of the ordinary. Indeed, little official concern was shown over the murder—even by Antonetti—beyond the fact that news of it would make recruitment more difficult, leading, in the governor-general's words, to "mediocre results" for future efforts. One official went so far as almost to justify Kondo's brutality by pointing out the "extreme difficulties" of recruiting in villages devoid of any "able men." At least some officials saw intimidation as part and parcel—even a necessary component—of successful recruitment.

———

AT BANGUI, MEN AND women from the far north were loaded onto river boats or barges for transport to Brazzaville. Albert Londres's description of the boats, in his signature spare, sharp prose, captured all that was wrong with the transport. The boats, many of which dated from the earliest days of the conquest in the preceding century, were ill designed for carrying human beings. In his account, Londres did not fail to invoke the slave ships of the Middle Passage. "In this country," he wrote, "the barges, not being made to transport men but only merchandise, had round backs. Three hundred by three hundred, four hundred by four hundred, the human cargo was packed below and above. The voyagers inside suffocated," he continued, "those outside could neither stand nor sit. What's more, having no foothold, each day—and the descent to Brazzaville lasted fifteen to twenty days—one or two of them slipped into the Shari, into the Sanga, or into the Congo. The barge continued." Once the boats arrived, Londres concluded, they carried fewer men than they had started with: of the original 300, only 260 were left, or maybe 280. And once on shore, the workers weren't even allowed to disembark because no one had planned to build a camp.

Londres's account was but one of various reports that arrived at the colonial minister's office and crossed Antonetti's desk. In late 1926, René Maran, in the *Journal du peuple,* reported on the deplorable conditions that men and women endured. He sent a transcript of the article to the colonial minister in an effort to promote what he called "a humane and rational colonization" of Equatorial Africa. In his article and his letter, Maran reported that the boats being used had been intended for nonhuman cargo; there was suitable room only for the ship's crew and a few travelers, not for hundreds of workers.

The trip caused constant misery for the men and women aboard. "If these passengers are dressed workers," he noted, "their clothing is burned by the embers that ceaselessly fly out the wood-burning boat's chimney." If they were naked, then the ashes burned their skin, forcing them to place an only "slightly protective" mat over their heads for as long as the wind blew ashes on them. Maran witnessed this firsthand on a trip north with workers who had finished their commitment and were being repatriated from the construction site to Bangui. "The majority of them were too poor to possess even the least bit of mat; they knocked away the embers with their hands after each burn." He added, "These burns are always painful; often remaining very sore." Some men and women did not survive the trip; having survived the ordeal of months on the construction site, they died one by one just as they neared home. Seven of the ninety-eight people on board Maran's boat died during the course of the eighteen-day trip. "It was pitiful."

Maran wrote to the minister in 1928, over a year after the publication of his article, because nothing had changed. He noted that the League of the Rights of Man had requested an investigation that had resulted in a "criminal amalgam of counter-truths" from officials in the colony. But conditions remained unimproved. Maran cited a letter he'd received from "a person very close to colonial questions" who'd also encountered boats carrying workers from the worksite back home. One boat had packed "like true cattle" some three hundred workers onto a single barge, leaving these "exhausted" workers in "a lamentable state"—"actual traveling cadavers, pushed together one against the other, of an unimaginable thinness, skeletal." They were only fed rice—"and what rice"—and some smoked fish. As a result, during the eleven days of travel, eight people died and one disappeared, his fate uninvestigated. "Never have I seen, for my part," Maran's correspondent had written, "sadder or more lamentable things." His correspondent had agreed with Maran: "*It seems that it's the same on every voyage.*"

Maran was not telling the ministry anything it did not already know. In August 1926, a military officer traveling through the colony had submitted his impressions of two boats he had encountered. Overcrowding was complicated by the presence of women who had to be housed in a makeshift hut made out of mats on the deck, where they "suffocated" in the heat. He added he didn't know how these people were able to survive a voyage of twelve to fifteen days. To improve the situation, however, the officer suggested not greater sympathy or regulation. Instead, he said, Africans needed discipline. He insisted that these Africans, unlike subjects elsewhere, were "lazy children" lacking civilization. They could not be administered; they had to be commanded: "they have to feel the European boss next to them, living with them, obliging them to work, keeping them from starving to death, visiting them, encouraging them, if need be punishing them."

In November 1927 a commission made up of administrators and barge company officials discussed the regulations needed to transport workers safely and effectively. But the meeting succeeded in only showing how regulations bordered on the Kafkaesque. For example, the commissioners agreed that a "water closet" should ideally be installed at the rear of each boat. But admitting that they could not expect ship owners to install water closets, they required them only in principle, not in reality. Likewise, stoves for cooking were also deemed necessary, though none had arrived yet from France. And while there was little to no form of cover on the boats, the commission deemed the boats too old to modify: "The protection issue," it noted grandly, "is thus considered regulated for this old model of boat." In short, if something was impossible to change, it was to be considered regulated; if boats couldn't be improved, then they were good enough.

Aside from amenities like toilets, cooked food, and protection from the sun and frequent rain, the commission's primary concern was how to pack as many workers on board as possible. It determined that, from a financial point of view, it would be ideal to fit five hundred workers on each boat. From an administrative perspective, 250 would be acceptable. But considering the dangers of overcrowding and discipline, the commission agreed, no more than one hundred should be attempted. The decision to carry one hundred workers made no mention that the same commission had determined that each of the forty-ton barges at their disposal could not safely carry more than seventy-five workers. In other words, even as the commission hammered out regulations, it took the fiction of their very efficacy as a given.

The attempt to carry as many workers as possible on each boat inevitably meant that the workers would spend many days or weeks of travel in ships' holds, which administrators deemed largely uninhabitable "for lack of air." Again, the insalubrity of holds, which as a rule became unbearably hot in the central African climate, did not deter administrators from regulating—and thus condoning—their use for human cargo. Their calculus for transporting seventy-five workers on a boat required that thirty-five of them be placed in the hold. The commission suggested that workers forced below deck should not be farther than five meters from the hatch and near no noxious products. The general aim of regulation was to ensure that workers, whether above deck or below, could expect one square meter each for the two-week voyage. The one-square-meter allotment, though, posed a major problem to the administration: as the commission report noted, offering one square meter per passenger was "going to have as a first result a considerable diminution of the workforce transported."

Regardless of what the commission put on paper or discussed at its meetings, little was done to improve overcrowding. In February 1928, months after the commission met, an administrator reported finding 208 workers pressed on board the *Dolisie*—a boat that Brazza had taken when he was conducting his investigation over twenty years earlier—so tightly they had no means of lying down. The Belgian ambassador—who was hardly one to criticize conditions in other colonies, considering how often his own country's possession in the Congo had come under scrutiny—complained to the French government of the "deplorable conditions" in which workers and their families were being transported. One boat was seen carrying six hundred Africans without any protection from the elements; something had to be done to avoid "such an intolerable situation."

The only response to these accusations that Brazzaville could muster was to deflect by way of correction. The boat the Belgian ambassador complained about, Antonetti insisted, was carrying only 591 passengers, not 600. He admitted that number was "excessive," though he didn't say how excessive. In his zeal to shift blame from the administration, Antonetti blamed the boat's captain—a Belgian national, he was quick to point out—for whipping the workers with a *chicotte*. What he did not admit was that the boat in question—the *Yvonne*—and its two barges, by the calculations of the commission, should have been carrying only about 160 workers, not 591.

Knowledge of terrible conditions and efforts at reform did little to effect change. A year and a half after the Belgian complaint, more mistreatment was linked to the *Colonel Klobb,* a very small steamboat suited for a handful of passengers, not for groups of workers. In the early 1930s a young Marcel Homet, who would later become a well-known and respected archaeologist and ethnologist, sailed on the *Klobb* and said that "men, women, and children would sleep stacked on the deck." After a short period of time on the boat, "the passengers, exhausted, no longer have human faces. They are dirty, their features drawn." In responding to inquiries about conditions on the boat, the captain of the *Klobb* replied bluntly, "I don't give a d—about workers, what interests me are passengers and freight." One official reprimanded the captain, insisting that "such language and lack of politesse" was unacceptable. Evidently, the misery of recruits packed on boats was less offensive than the captain's vulgarity.

For years on end, then, the men and women recruited in the north of the colony faced about two weeks of travel on an open, winding river, with little protection from the elements. Annual average daily high temperatures for the region of Ubangi-Shari, from where many boats departed for the south, ranged from the mid-80s to the mid-90s Fahrenheit. Throughout the year, relative humidity levels in this part of central Africa consistently peak at over 90 percent. Thunderstorms were a regular feature, with most days in the wet season (which lasts more than six months of the year) experiencing rain. During daylight, men and women spent hours in heat, either on the unshaded deck or in a windowless hold, punctuated by heavy rains. The rain might well have offered refreshment at first, but it also left recruits to spend nights in the wet and cold.

Added to the natural conditions were man-made crises. In the details of regulations, it was easy to lose sight of the human experience of boat travel. Crowded with standing men and women, boats were sullied with the stench and filth of urine, feces, and vomit. The lack of toilets on a boat carrying anywhere from two hundred to six hundred passengers meant that men and women, pushed together to the point of immobility, had little choice but to soil their own space. Many voyagers were sick with intestinal illnesses that caused intense diarrhea. Boats did not have doctors on board to serve recruits. At best, the sick stood at the back of the boat so as not to contaminate the healthy. In one documented case, separation did not stop contagion, which caused many on board to sicken or die. For the sick and the healthy alike, the sanitary conditions alone were appalling.

If illness weakened many, so too did hunger. The food ration of rice and dried fish provided less food than concentration camp inmates received in Nazi-occupied Europe. Had the African recruits on board been fed a bowl of rice three times a day, a generous estimate considering reports of shortages, and a serving of dried fish, their total daily caloric intake would have been about 1,200 calories—comparable to the rations at Auschwitz and about 30 percent lower than those at Buchenwald during the Second World War. The basal metabolism of an adult male at rest requires 1,500 calories daily. The recruits transported on boats obviously did not face the heavy labor that concentration camp inmates did, at least not until they arrived at the construction site. But considering the overcrowding, they were also certainly not at rest. Many spent days on end standing, bodies pressed up against one another, sitting or sleeping in short shifts. Such fatiguing conditions left them in a state of semistarvation—a situation that did not always change on the construction site.

The continued debates among politicians, journalists, and administrators showed that few significant improvements were made before, at the very earliest, the end of the 1920s. By the administration's own count, then, at least 45,000 travelers steamed slowly along in a fluvial world of sickness, hunger, fatigue, and despair. There is little mystery why so many arrived, if they survived the trip, in a deplorable—or as more than one witness noted, "skeletal"— condition. Some recruits could not tolerate the transport, choosing instead to jump into the river, where crocodiles were waiting.

CHAPTER 5

"THE MAYOMBE DOESN'T
WANT US"

You build your kingdom upon corpses.

RENÉ MARAN

FOR THE MEN, WOMEN, and children who survived the hundreds
of kilometers of marching, the sun and rain, the hunger, and the
overcrowded barges, arrival at the worksites of the Congo-Océan
provided little respite. Recruits from the north were housed for a
period in Brazzaville. The Congo-Océan had helped transform the
colonial capital, increasing both its African and its European populations, though it remained, in Marcel Homet's assessment, "only
a big, very ill-equipped village." He counted just over a thousand
inhabitants, who all lived without reliable water, sewage, or electric-
ity. In truth, Brazzaville was more like a collection of disconnected
and racially segegrated villages, including a small central European
district, a Catholic mission, and the two main African villages of
Poto-Poto and Bacongo. Once the railroad was complete, the city
would become a key transit point for goods coming from upriver. In
the decade before that, it was the place where recruits, over a thou-

sand at a time, were held in camps and introduced to what passed for French "civilization" in Equatorial Africa.

In Brazzaville, recruits had medical inspections and time to rest from their travels, to acclimatize, and to be trained to be productive workers on the construction site. Then began the process of remaking villagers into laborers. As officials imagined it, this was the moment to train workers in clearing, terracing, and sanitation, in how to use a shovel and how to push a minecart. Recruits were meant to be taught French notions of hygiene, to feed themselves, and to keep their bunks. Those deemed "unfit" or "weak" were given medical treatment or reassigned to less demanding jobs. Those who were healthy were sent to their new assignments, to camps that lay two weeks' to a month's hike from Brazzaville.

Working and living conditions were taxing and dangerous all along the line. But one stretch of the railway more than any other came to represent—in the minds of workers as much as in the pages of books, articles, and reports—the extreme physical and emotional challenges of the project. The Mayombe is a low, densely forested mountain range that stretches southward from Gabon through Middle Congo into what is now northern Angola. The mountains of the Mayombe are not terribly high—few peaks are more than three thousand feet—but the tree-lined ridges and precipitous flanks capture cool air and cast much of the region into darkness, even in the midafternoon. The Congo-Océan passed through about seventy miles of the Mayombe. The entire railroad construction site was measured and mapped by the kilometers of each point from the end of the line at Pointe-Noire; the Mayombe stretched roughly from Kilometer 60 to Kilometer 170. To the west of the Mayombe lay the coastal plain that ended at the Atlantic; to the east were the rolling hills of the plateau that led to Brazzaville.

While accounts of the railway could differ dramatically, almost everyone who visited or worked in the Mayombe agreed that it

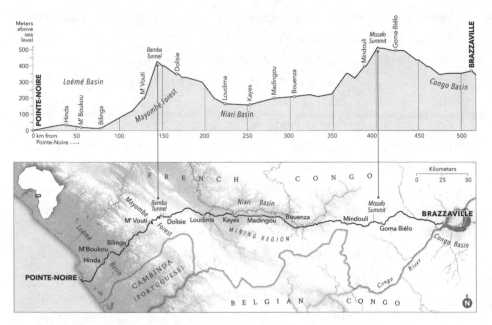

Route and corresponding altitude of the Congo-Océan railroad.

was an utterly inhospitable place to do much of anything, let alone build a railroad. In the 1930s, Marcel Sauvage, a journalist, poet, and essayist, penned an evocative description of the Mayombe in *Les Secrets de l'Afrique noire.* Sauvage went to the Mayombe to see for himself the project that had, in his estimation, already cost fifty thousand workers their lives for "the useless establishment of a little railway line that could have gone elsewhere." But it was the forest—a world unto itself, a primordial realm plagued by an unrelenting climate and vibrant flora and fauna—that fired his imagination. The "low, sick, violent sky, lacerated by lightning," he wrote authoritatively, "dumps between seven and thirteen meters of water annually"—the height of a four-story building. It was a striking image, if wholly overstated; annual rainfall in the

Mayombe is less than seven feet—still an extraordinary amount. Even when the sky did not open up with rain, he noted, the forest choked man and beast alike; it was a "hell of humidity where, apart from several outcasts of [African] humanity, only prodigious gorillas survive."

"The virgin forest of the Equator," Sauvage continued, "constitutes the triumph of vegetation, in an atmosphere of steam and fever"—a scene that was nothing short of "ferocious and shivering." Trees reached to the sky, creating a "leafy vault" above and a "mattress of rot" below. "From dawn," he wrote, "we were attacked by clouds of flies, and in the evening, as soon as my lamp was lit, a thick swarm formed, an aerial purée of *niamas*, mosquitos, into which the lamp and I completely disappeared after about five minutes." Such tribulations took unspeakable tolls on mind and body. For part of his trip, Sauvage traveled with a civil servant who developed a skin disease that rendered him "no more than a skeletal generator of pus." Here in the low mountains of Equatorial Africa were plagues of darkness, locusts, and boils. "You lose the measure and the notion of human realities," Sauvage observed. "You sink into a dream that constantly turns into a nightmare." The Mayombe, in short, was "the forest without joy."

Sauvage's prose certainly had its purple flourishes, but his was hardly the only account of the Mayombe to revel in its miserable and otherworldly characteristics. Suffocation was a common theme in other travelers' accounts. Gabrielle Vassal wrote that, on the construction site, "a deadly weight seems to squeeze all life out of brain and body." Albert Londres remarked that he had never seen such trees; as soon as he and his companions strayed from the trail, "the forest closed around us like a tunnel." Robert Poulaine echoed this sense of being enveloped. He described the canopy of the forest as a "dark vault," with "gigantic trees interlaced with creepers large like

In the Mayombe French Congo, postcard, c. 1910.

bushes, undergrowth infested with thorny plants." The atmosphere, Poulaine continued, was "heavy and fatally debilitating to the lungs of whites and blacks" alike. "And that's not all!" The forest was a "vegetal barrier" put up by "a nature rebellious to human progress."

Photographs of the forest offer glimpses of its intimidating majesty. A number of photographers and postcard makers tried to capture the Mayombe's impenetrability, even occasionally in stereoscope. The results of shots straight into the bush were flat, dark, dense images that struggled to capture the multistoried forest. At best, they offered a mere evocation of the towering evergreens and palms, the foliage of herbaceous and epiphytic plants, and the steep mountainsides that left not a horizon in sight. Where paths had been cut for road or rail, photographs revealed canyons of foliage, with towering trunks and dense canopies forming walls along the edges of the clearing. They hinted in sepia tones at writers' allusions to Bible and epic. The Mayombe was mile upon mile of "Dantesque chaos," of "Apocalypse." The rails would have to cross it if Brazzaville were to be linked to the sea.

Europeans and Africans alike knew of the Mayombe's dangers. In 1925 one official noted that the route through the Mayombe "brings the railroad . . . into a deserted region, insalubrious to the point that the former inhabitants have disappeared, and the European personnel as well as the native labor will have much to suffer from." Another reported that the "chaotic configuration of the ground, with steep hills separated by deep ravines," and its adverse impact on health, gave the region a menacing reputation among the indigenous population. The hostility of the climate led African workers to coin an expression as simple as it was accurate: "The land of the Mayombe doesn't want us."

Early plans for the railroad had bypassed the most treacherous parts of the Mayombe. But changes, including the decision to build the main coastal port at Pointe-Noire and a desire to shorten the route, drew the final path right through the region. The Mayombe thus became the primary center of effort. It was where the engineering feat was at its most bravura: unstable soil, precipices and chasms, waterways and erosion, meant that the Batignolles had to construct hundreds of small bridges, massive viaducts, and a 1,700-meter-long tunnel under Mount Bemba, the longest in Equatorial Africa.

Until 1928, when part of a service road was completed, all food and supplies had to be carried into the Mayombe, along "goat paths" if any at all, on African porters' backs or heads. The construction effort in the mountains was intensive. One engineer estimated that it would require more than 10 million individual workdays to complete the passage through the region, or about 107,000 days per kilometer of track. To complete this stretch, the Batignolles would have to rely overwhelmingly on a single source: more and more men.

By the late 1920s, to the railroad's critics, the Mayombe had come to represent the folly of the project. In 1929 a French Communist deputy in the National Assembly in Paris explained the deadliness of the Congo-Océan by simply pointing to where it was being built. The Mayombe, he noted, was "an unhealthy, humid forest which hardly sees the sun and where the work is extremely tough"; his colleagues on the far left cheered, *"Très bien! très bien!"* in agreement. The deputy's reasoning might have been overly reductive, but it contained certain well-known truths. Tens of thousands of men and women passed through the Mayombe; many thousands of them never went home.

———

WHAT, THEN, WAS LIFE like in the Mayombe for the workers who arrived there, usually after a long and exhausting journey? A variety of accounts, from both Europeans and Africans, reveal attitudes, complaints, and acts of resistance that provide insight into workers' experiences and concerns. And many of these same documents offer trenchant indications of just how unprepared the colonial administration was to build a railroad in Equatorial Africa. It left the recruitment of laborers in Ubangi-Shari and Chad to often understaffed and unsupervised officials in distant fields. But on the construction site itself, there was no handicap of distance. The governor-general could and did visit the *chantiers* with some regularity. Reports of mistreatment, poor living and working conditions, and extraordinary mortality crossed his desk regularly. Workers suffered immensely on the Congo-Océan, especially in the Mayombe. They were not victims of a well-oiled machine of colonial brutality; more often than not, they were incredulous witnesses to European hubris, casual callousness, and massive organizational failure.

For men and women who had infrequently traveled far from their kin groups and ancestral homes, a trip of many hundreds, or even thousands, of miles left them exhausted, disoriented, and scared. Combined with the general angst of displacement, many workers were well aware of the reputation of the railway. From the beginning of their recruitment, men arriving in Brazzaville from the north regularly refused to continue to the construction site, saying that death awaited them there, if not immediately, then after an agonizing period of mistreatment. The fear of death intermingled with their sense of isolation from their communities. They feared that if they died, they would be denied the proper funeral rituals that would assure peace to their souls.

Added to such challenges were the stresses caused by the Mayombe's darkness and unforgiving weather. After 1925 tens of thousands of recruits arrived from much drier climates, primarily deserts and savannahs to the north, and were unaccustomed to dense and damp rain forests. In this radically different environment, they were forced to adapt to a completely foreign way of life. They regularly felt adrift, removed from their villages, driven by strangers to a location that few could even imagine. Officials tried to sort the men by their ethnic or linguistic group, with mixed success. Regardless, workers found themselves separated from the close-knit communities they had experienced at home. Unable to communicate effectively, exhausted by months spent in transit, often hungry or starving and homesick, they now inhabited a world unlike any they'd ever encountered. Doctors and officials cited homesickness as a constant concern for morale. Workers found themselves, as one report put it, *déracinés, dépaysés, démoralisés*—"uprooted, exiled, demoralized."

Workers' sense of isolation and disorientation was shaped by language as well as by distance. The lingua franca of the Congo-Océan was French, but very few workers had any knowledge of it. The imposition of French was an immediate assault on their humanity. In reflecting on his imprisonment by the Nazis, the Italian writer Primo Levi observed that knowing the language of one's captors represented a watershed: those who understood German could have a "semblance of a human relationship." For those who did not understand, the interaction was full of terror. "An order that had been pronounced in the calm voice of a man who knows he will be obeyed," Levi observed, "was repeated word for word in a loud, angry voice, then screamed at the top of his lungs as if he were addressing a deaf person or indeed a domestic animal." For Levi, the ability to communicate was a "necessary and sufficient mechanism for man to be man."

To many European minds, the inability of African workers to speak French only supported the belief that they were docile, stupid, or less than human. European overseers regularly complained that their workers did not understand them—as the governor-general put it, workers spoke "thirty different dialects—nearly impossible to identify." (Not that many Europeans tried.) The inability to communicate led to frustration, anger, and often violence. Compounding the communication problem, not all European employees were competent in French: the white *surveillants,* or foremen, who directed daily tasks on the site often came from Italy, Russia, Greece, Portugal, Poland, and elsewhere. In 1925, for example, only three of eleven Europeans along one twenty-kilometer stretch of the construction site were deemed to speak decent to good French; five spoke hardly any at all. As Albert Londres quipped, "It's no longer the Congo-Océan, but the Congo-Babel."

White *surveillants,* then, barked grammatically broken orders to their *capitas,* the African overseers who prodded the laborers to work. *Capitas* often spoke neither French nor the languages of those under their command, so they repeated orders as best they could, often at volume. The progression from talking to yelling to hitting worked its way down the chain of command. Londres saw an Italian overseer scream at his workers, "Bastards! Pigs!" His *capitas* followed his lead, repeating the insults "like an echo." As he described it, the *capitas* hit the workers, and the workers hit the rocks. Across the worksite, "awesome disorder" reigned; all one heard was yelling.

If being transported to a strange and dangerous land left recruits feeling lost, isolated, and abused, the administration took steps to bring regularity to their lives. In 1924 the Colonial Ministry issued orders requiring all aspects of workers' lives to be subject to "careful regulation." Upon his arrival, Antonetti created a Labor Service to oversee workers' health, to hear their complaints, and to study living conditions with an eye to reform and innovation. As one administrator put it, its goal was that of "treating workers with more benevolence and humanity." With thousands of new recruits arriving from hundreds of miles away, the Labor Service was taxed with the job of trying to understand workers' lives, from their emotional and physical health to the ways they spent their hours, be it working or relaxing. While ultimately it would not be terribly effective at greatly improving the lives of workers, it would produce reports that made plain the administration's continued challenges to live up to its regulations.

As with many aspects of the Congo-Océan, the organization of the construction site looked decent on paper. Housing was to be designed according to the needs prescribed by doctors to guarantee health and hygiene. Food, of sufficient quantity and appropriate quality, was to be provided. And when possible, workers' morale was to be boosted—interactions with family, dances, and sports were "particularly encouraged." Pay was minimal—until 1930, it amounted to less than two francs a day for men, a few centimes more for *capitas,* and less than a franc per day for women. By contrast, white construction workers and overseers could make 1,000 to 2,000 francs per month. Pay for workers was also often withheld, certainly if they became sick or injured, but also if their overseers deemed their effort insufficient—a practice documented by inspectors as early as 1924 and as late as 1931.

The administration aimed to offer healthy, relatively comfortable living and working conditions for its workers. In practice, life in the camps throughout the entire period of construction was an awkward balance of group discipline and self-reliance that revealed the tenuous organization of the project. Officials sought an environment of military organization. Instead, camps could resemble prisons as much as the housing of allegedly free labor. Camps were guarded by *miliciens* or *gendarmes,* armed African men known for their brutality, who were often of different ethnic groups from those they guarded. Whether standing for roll call, reporting for medical inspection, or making the trek to and from the construction site, workers were usually "pushed" by armed guards.

Housing was often lacking or in poor condition, especially before the 1930s. As late as 1928, only one camp in the entire Mayombe—

the one in M'Vouti, one of the larger centers on the railroad—was deemed "well organized" by the administration's own inspectors. The challenge for the administration was to construct clean, adequate housing for workers along a railroad where the location of the worksite was constantly moving. Housing tended to be best in the main transfer and administrative points along the line, especially Pointe-Noire, M'Vouti, the copper mining center of Mindouli, and Brazzaville. The Brazzaville camp, for example, had housing with circular verandas, high ceilings, and spacious rooms with individual beds, as well as a store, a washhouse, and even a pool. But most of the camps of the Mayombe were far less attractive, often failing even to provide decent protection from the cold and rain.

Progress along the line was certainly slow, but even so, the center of construction moved twelve to twenty kilometers per year, either rendering housing locations inconvenient within a matter of months or forcing men to spend a considerable part of their day walking to and from sites. Along some parts of the line, camps were constructed every two kilometers, a fact that one administrator suggested was a sign of "our feelings of humanity" since it meant workers did not have far to walk. But in reality the difficult terrain of the Mayombe made walking even short distances onerous and slow; in some areas it could take an hour to cover a kilometer or two. Considering that workers returned to their camps for lunch, even short distances could mean they spent several hours a day walking.

Throughout the 1920s, most of the housing consisted of large common barracks or huts built of sticks, mud, and foliage, often in some variety of wattle and daub. Large barracks could be subdivided into rooms for as many as fifty to sixty men, as well as the few wives who joined their husbands. Some were smaller, housing eight to ten. Workers were assigned either a small single bed or, more commonly, a shared bedstead upon which a number of men slept. Interiors were cramped, dark, damp, and cold; many buildings had few windows and no source of light. One rather blunt medical inspector referred to workers' housing as an "extreme euphemism for architecture."

For their part, the men and women who had to live in these structures were dubious, and rightly so, that the French knew how to construct shelters properly. They believed the buildings attracted ticks, jiggers, and parasites—concerns that were, in fact, well-founded. Robert Poulaine deemed the housing even more danger-ous, lamenting the misery and death it caused. The buildings were, he wrote, "straw huts where water penetrates and remains, where nothing is dry, where malaria follows you, as well as all the compli-cations of hematuria, bile, not to mention dysentery, fatal to Whites and Blacks alike."

Barracks usually had dirt or mud floors, even though inspec-tors recognized the superiority of cement. Roofs made of foliage rotted in the humidity and rain, leaving gaping holes and offering little protection from wind and cold. Rooms had no light: in the hours after work and before bed, workers ate and talked in profound darkness. At best, they fashioned torches from straw, a dangerous fire risk. Few barracks had stoves or heaters of any kind. Workers would gather in groups of four or five around a small fire to keep warm—they awoke in the morning "smoky and numb." In some cases, workers used parts of the walls and roofs to burn as fuel.

The filth and stench of the housing must have been overwhelm-ing. Latrines were in short supply, especially in the early years. In camps without them, men were encouraged to bury their feces, though it seems few did, leaving the area around and even inside the buildings soiled with urine and excrement. Barracks were often the final refuge of the sick and dying. The bodies of the dead were regularly left in huts or the neighboring bush for days before being removed. As most camps lacked incinerators, garbage piled up. Daily housework was left to the women living in the camps and to men too weak to work. The bulk of cleaning and disinfecting was meant to be done by workers on Sundays, the day of rest. Perhaps not surprisingly, little was done then, leaving camps in a derelict, unhygienic state. The stench, vermin, and fear of disease of the administration's housing encouraged many men to sleep outdoors. The unpredictable natural elements were preferable to what awaited them inside.

African workers had little protection against the cold and damp, an ever-present tribulation in the Mayombe. Most worked barefoot and unclothed, using their single camp-issued bedcover as a loincloth or poncho. Doing so, however, rendered the bedcover too wet and dirty for use in camp. Others fashioned clothing out of old cloth sacs. Photographs of work teams revealed that workers regularly made garments from whatever they could. In 1926 the colonial minister ordered the provisions of a uniform upon recruitment. But administrators balked at the idea, saying that Africans would refuse to wear clothing. Two years later new regulations promised two khaki garments at the moment of recruitment, with replacements provided every six months thereafter. In practice, where workers had uniforms on the construction site, and distribution was limited, they were often in tatters.

Food and water were also, in theory, regulated. Upon arrival at camp, workers were provided a mess tin and a spoon. In 1926 officials discussed supplying men with lunchboxes, but no local merchant could be found to provide them. Not until 1928, some seven years after the start of the building of the railroad, did an administrator suggest that men and women receive a can from which to drink water on the construction site, as well as basic kitchen tools to cook their meals. Daily food rations were also regulated. That said, food was regularly in short supply, often inedible, because it was either rancid or not appealing to the palates of people from radically different climates, and was often lacking in the vitamins and calories needed to keep men and women healthy.

Camps lacked basic water infrastructure, including barrels to capture clean rainwater. Even in camps close to Pointe-Noire, workers were left to drink polluted water, much to the detriment of their health. "I have seen them," one health inspector reported in 1928, "drink from green and stagnant water." Not until 1929 did the administration improve methods of collecting and distributing water, and even then it was not universally accessible. In late 1931 a camp that housed hundreds of men was found to possess only one faucet. Because the faucet drained much of the camp's limited water supply, workers were not allowed to bathe; but there did appear to be sufficient water to wash the equipment used on the railroad.

Having attempted—albeit with very mixed success—to provide for every worker's basic needs, from clothing and housing to food and water, the administration also expressed a desire to recreate some of the more intangible comforts of home on the Congo-Océan. In 1925 Governor-General Antonetti ordered that "nothing will be neglected" to assure that men on the construction site enjoy "the same pleasures, games, dances, etc.," that they had in their home villages. It was here that the presence of women played a key role.

The lives of women on the Congo-Océan are remarkably absent from the colonial archives, even relative to the thousands of men who are themselves often absent or silenced. Women's experiences, feelings, and opinions were simply not of concern to an administrative machine preoccupied by finding men to do manual labor. Officials did not even keep reliable statistics of the number of women present; regulations on female recruits—their number, the conditions under which they would come—were unclear throughout the construction. In 1926, during a promotional tour of the north, Antonetti encouraged much-sought-after Sara recruits to bring their wives with them. This was likely aimed more at winning over male recruits than at bringing women per se.

That said, women not only were present on the Congo-Océan, but they were also valued by the colonial state for the contribution they made to the construction. If wives were ideal, recruitment also clearly netted many unmarried women, and indeed, many wives became widows along the way. The ranks of female recruits were also supplemented by women who lived in villages around the construction site, who either filled short-term jobs or fulfilled their obligatory colonial labor tax through work on the railroad. A 1925 photograph of a woman from M'Boukou, on the railroad route, makes clear that women did this work in addition to their myriad responsibilities as mothers, wives, and daughters.

Women fulfilled many of the essential jobs that they had done at home, and, paid a fraction of what male workers received, they did so at a price enticingly affordable to the colony. All workers on the construction site were expected to prepare their own food, which could prove difficult for men, especially when they had to walk long distances to and from camp for the midday meal. Women collected wood, built fires, and prepared meals for small groups of men. The administration hoped that women would also plant gardens or gather edible plants in the forest to supplement diets. In addition to cooking, they carried water, cleaned the camps, and tended to the sick and wounded. With grasses, reeds, and wood, they could make mats, woven items, and tools to be used around the camps or to repair huts. In more relaxed moments, women participated in the rituals and pastimes of camp life.

Women were brought to the construction site to help recreate for male workers a sense of the normal rhythms of village life. While it remained unspoken in official paperwork, sex was undoubtedly part of this colonial calculation. The preference for married women might have reflected, at least on paper, an official effort to control sex. Some camps were meant to provide housing specifically for married workers. But women, married or not, were also clearly seen as servants of both African and white men's sexual desires. French prejudices had long considered African women to be sexually available, an assumption that did not change on the railroad. In his fictionalized account of his work on the Congo-Océan, Michel Romano (who published under the pseudonym R. O. Manot), wrote of the "bewildering ease of negresses" which enabled one aggressive white acquaintance to have sex with any woman he encountered "no matter where." The fact that women represented about 10 percent of the population on the line made them constant targets of male harassment.

Regardless of what the administration expected women's sexuality would bring to the camps, the results were far from uniformly positive. Workers had conflicts over women, some of which turned violent. Liaisons between men and women, no doubt further complicated if white men were involved, spawned resentments and vendettas. Prostitution became a part of the camp economy; women were exploited, abused, and fought over. On at least one occasion widows were linked to an outbreak of venereal disease. On other occasions, married men sold their wives' sexual favors to other workers for cash. The exchange of wages, women's refusal and mistreatment, and men's feelings of being cheated flared tempers; verbal disputes could turn to physical fights, sometimes forcing the administration's involvement. As marriage and sexuality shaped people's kinship networks and personal identities, attempts to recreate the gender dynamics of home on the Congo-Océan likely heightened men and women's sense of isolation rather than alleviating their homesickness.

Efforts to replicate some of the experiences of home ultimately failed to improve morale. In 1928 an experienced doctor and civil servant, Dr. Lasnet, inspected a number of camps in the Mayombe on Sundays, the workers' day off. He found "neither drum, nor merriment" in the camps he visited. "Workers give the impression

A woman of M'Boukou, c. 1925.

of weariness and boredom"; morale was not high. Dr. Lasnet suggested that the administration give some thought to making Sundays more special, maybe through the distribution of kola nuts or a bit of tobacco. Or perhaps, he continued, the French could open little shops to sell inexpensive trinkets to the workers that could be both a distraction and teach workers the value of money. Even relaxation and entertainment were controlled and aimed at instilling the habits of regimented, industrial workers. Alcohol, one possible form of release, was strictly prohibited; high fines and even imprisonment faced anyone who tried to sell it.

Instead, officials promoted exercise and athletic competitions that could raise spirits; little thought, it seems, was given to whether men and boys who did physically exhausting labor nine to ten hours a day, six days a week would want more exertion on their days off. It was, in the end, moot. Sports and games were out of the question because the administration had no one to organize such activities. There was "no moral assistance, no encouragement," Dr. Lasnet noted, "to put a cheerful note in their sadness."

―――――

HEAVY LABOR ON THE construction site was undertaken almost exclusively by men. Days started early, when the forest was cool and dark. Men awoke, reported to roll call, received their work assignments, and if physically able, began the day. They could be assigned to any number of jobs that the site managers (*chefs de chantier*) deemed necessary. Some were sent to work on major projects, under the direction of white *surveillants*, ranging from digging trenches and terracing to cutting stone and laying track. Other teams were assigned to *tâcherons*, white men who were hired usually for short periods of time to complete small projects. Whatever their assignment would entail, workers knew they faced a long day. Regulations called for ten hours of labor per day until 1927, when a nine-hour day was implemented. Work hours were sometimes shorter but often longer; a common complaint of workers was a lack of consistency of expectations. Officials noted that many workers toiled "at the pleasure" of their overseers and were sent back to camp only when the boss said so.

What did work actually entail for men on the construction site? More than anything else, men on the construction site had to clear brush and move dirt and stone. The quantity of foliage and earth that needed to be moved was staggering. First, a tropical forest had to be leveled and cleared. From the lowest creeper to the most towering tree, bush had to be cut and cleared, root systems excised. Once the forest was cleared, the topography could be reshaped. On one ten-kilometer stretch of the construction, Kilometer 95 to 105, a single *meter* of laid rail required an average removal of more than 12 cubic meters of dirt and 29 cubic meters of rock. To put this in more contemporary terms, each meter of track required, on average, the removal of about four and a half dump-truck loads of earth. For the ten-kilometer stretch as a whole, that would mean about 45,000 dump-truck loads of dirt had to be moved.

While the total quantity of earth moved is mindboggling in the abstract, for the worker the toil was monotonous and never-ending. Day in and day out, men went to and from their worksites to chip away at the monumental edifice of the Mayombe. In his 1987 novel *Le Feu des origines*, Emmanuel Dongala's character, Mankunku, hammers rocks all day long, making boulders into gravel. His hands are blistered and throbbing, and the clinking of stone rings in his head long after he's stopped working. Thousands of men were thus employed to turn boulders into stones, rocks into gravel. Stones would be placed in a train car, on a barrow, or on a man's head or back and be transported to other points on the railway line, where they would be shaped into building blocks for walls, bridges, or drainage canals. Gravel was shifted and moved to beds where the track would be laid.

It is difficult to quantify the effort that went into building the Congo-Océan, but it is clear that laying a railway through the dense forest and on the unstable land of the Mayombe posed formidable challenges for engineers and laborers alike. Once the landscape was cleared and transformed, construction began. Workers dug trenches and terraced mountains and built massive retaining walls. Through the Mayombe, the serpentine route wound around hillsides and rock outcroppings. Across one stretch of less than ten miles, between M'Vouti and the foot of the highest point, Mount Bamba, the tracks turned 197 times.

Where the tracks were to cross a river or gorge, men erected stonework foundations and supports upon which scaffolds were built to pour the cement of the bridges and viaducts. Where hillsides could not be flattened, they dug tunnels. In one major section of the Mayombe, from Kilometers 97 to 143, the Batignolles oversaw the construction of thirty-six major viaducts, seventy-three smaller bridges, twelve massive retaining walls, and thousands of meters of trenches and tunnels. The multiple arches of the largest of these viaducts loomed dozens of feet above rivers below.

Remarkably, the transformation of the landscape, as well as the construction of walls and bridges, was accomplished almost entirely by African hands—not metaphorically but in reality. Laborers cleared, dug, bridged, bored, leveled, and terraced countless cubic meters of earth without the substantive use of modern tools. The Batignolles claimed repeatedly that the lack of mechanization on the Congo-Océan was due to practical considerations, namely, the difficulty of shipping and moving equipment into the Mayombe. But the company's shifting attitude toward using modern tools was dictated by financial incentives more than any other factor. According to a 1925 agreement, the French administration remunerated the Batignolles based on the work completed, encouraging the company to keep costs as low as possible to maximize its own profits. Since the administration also provided labor at bargain basement prices (the state charged the company only a fraction of its actual costs), the Batignolles relied almost exclusively on human power, introducing minimal mechanization to the worksite.

The governor-general's office was not happy with this arrangement. Workers often had no tools for clearing forests or removing stone—no winches, cranes, drills, or saws. For the pouring of cement—from which many of the bridges and tunnels were constructed—there were no cement mixers and no rebar to reinforce the concrete. Nor did the Batignolles hire enough skilled workers to oversee the more complicated projects of masonry, tunnel digging, and bridge building. Antonetti did not hesitate to tell Marcel Rouberol that he was dissatisfied, accusing him of trying to build the railroad "almost without tools . . . and counting entirely on the docility and consent of the Congolese blacks."

*Consent* often came in the form of men being driven like beasts of burden. During his visit in 1926, on the outskirts of M'Vouti, Albert Londres came across men hauling a fallen tree without ropes or belts: there were but "the hands of blacks" to lift and drag. A *milicien*—possessed "by the demon of folly"—counted in a thick accent, "Wan! Tow! Thray!," as the men heaved. He stood on the tree that the men were struggling to move. A *capita* then beat the men with kicks and punches. Londres reported that four men left the tree with crushed fingers; two others had faces bloodied by the guard's whip; yet another had an injury on his neck. The tree had not budged.

Londres had seen railroad construction sites before, and he was well aware of the implements and materials usually used. "Here, it's only the black man! The black man replaces the machine, the truck, the crane," he wrote, "why not explosives too?" With his typical sarcasm, Londres added, "I discovered on these worksites important instruments: the hammer and the crowbar. In the Mayombe, we drill tunnels with *a* hammer and *a* crowbar!" Official reports corroborated the journalist's account. One onerous task that men were required to do was clear foliage, removing the massive trunks, twisted vines, and expansive roots of the equatorial forest. In 1925 one official noted that three hundred men had been given only fifteen axes and several machetes to clear a broad swath through an area with giant trees, bushes, and vines. The rest was left to be done with hands and backs, driven on by *capitas'* cries.

But the Batignolles' methods had their defenders. European commentators even defended the lack of mechanization by blaming workers who were incapable of understanding modern means of construction. African workers' "methods were primitive in the extreme," Gabrielle Vassal wrote, "engineers had evidently great difficulty in obtaining the least effort from them." Vassal went so far as to suggest that workers' practice of carrying dirt "in small baskets on their heads" was a sign of French engineers adapting to African ways of work. With a quick turn of the pen, she explained away the lack of tools by reference to the allegedly primitive and lazy ways of the workers, rather than to the penny-pinching ways of an international construction firm.

Such arguments, however, did not fly with critics in Paris. In the Chamber of Deputies, politicians condemned the original agreement with the Batignolles that made workers expendable; according to one deputy, when one man fell, he was simply "replaced automatically and without charge from the administration." A socialist deputy linked the lack of machinery to the deaths of "thousands of blacks . . . at least twenty thousand." He told his colleagues that there was no excuse for the lack of mechanization: "Even in a country of primitive civilization, a public works contractor must take care to avoid, with the use of machines, the loss of human lives."

The debates in Paris encouraged Antonetti to demand changes from the Batignolles. The governor-general's office repeatedly pushed the company to adopt basic equipment, such as winches and stone crushers, but faced constant resistance and excuses. Finally in 1930, the labor arrangement between the state and the firm was renegotiated, including new financial incentives to encourage the Batignolles to increase mechanization. With a new bottom line, the Batignolles responded accordingly. But by that time, much of the most difficult work in the Mayombe had been completed.

———

CLEARING, DIGGING, AND BUILDING were obviously essential to the construction of the Congo-Océan. But they would have been unthinkable were it not for thousands of men and women who performed the thankless tasks of hauling heavy loads as porters. Porterage, which had a notorious reputation in equatorial Africa stretching back to the earliest days of conquest, was absolutely essential to the construction. It was estimated that one porter was needed for every worker on the construction site. And it is not difficult to see why: for years, the vast bulk of all the supplies, from building materials to food and water, had to be carried from the ends of the line to the interior. The distances that had to be covered were significant. In 1924 a large portion of supplies arrived from over a hundred miles away along tortuous paths.

Upon his arrival in 1925, Antonetti pushed to build a service road alongside the railroad, ostensibly to help deliver goods and men by truck to parts of the Mayombe that could otherwise be reached only on foot. In making his argument, the governor-general became a vocal critic of the reliance on porters. "It's horribly costly, inhumane," he wrote to the colonial minister, "for porterage is by far the natives' most painful, most dreaded work and it causes the most illness." In his communiciations with the Batignolles, he encouraged the company to use porters as little as possible, for only short trips, and to limit all burdens to no more than 25 kilos (about 55 pounds). To fail to do so, he insisted, would be "exhausting and repugnant to the men."

Marcel Rouberol, at the Batignolles, brushed off Antonetti with two blunt truths: nothing in the agreement with the company prohibited using porters, and it was the only means of transport in the region. So porters were used. The practice did have more verbose and enthusiastic defenders, as well—including Antonetti, when it suited him. Indeed, belief in the Congo-Océan required an acceptance of porterage. An administrative note on the subject, probably written to help work out a rhetorical defense of porterage, stressed that Africans were always carrying all sorts of things, from goods to children, on their backs or heads, even while cultivating their land and dancing to the tamtam. Porterage was necessary, the note rationalized quite simply, for a "colonization capable of transforming a savage country into a civilized country."

If the French were introducing the people of Equatorial Africa to civilization, porterage was a perverse way to do it. Nearly everyone, from administrator to worker, acknowledged that porterage was the most miserable and dangerous of all assignments. This had been the case for porters long before the Congo-Océan project began and in parts of the colony far removed from the construction site. André Gide, in condemning the practice among concessionary companies, cited reports from the early 1900s of entire villages dispersing. In Ubangi-Shari, according to a 1902 report, one would "abandon his tribe, his village, his family and his crops, go live in the bush like a hunted beast in order to flee the recruiter." Villagers preferred everything—"even death"—to porterage.

If porterage was a fate worse than death in the relatively flat Ubangi, the broken topography of the railroad made it all the more odious. The trails that porters followed were often narrow, wet, and slippery, crossing "chaotic and wooded terrain" through the forest. Bearing heavy burdens on their shoulders, backs, or heads, porters were forced to ascend and descend steep passages and cross rocky massifs, sometimes using ladders. Even along the construction site, as trees were being cleared, porters had to navigate the crisscrossing branches and enormous thronelike trunks that had been cut but not cleared away. Under such circumstances, it could take an hour or more to progress a single kilometer.

Although porters were not, by regulation, meant to carry more than 25 kilos, they were forced to do so regularly. One report from late 1926 suggested that a 95-kilo load (over 200 pounds) was common for two men to lug together. Staple burdens of the porter's trail were barrels of cement that came in containers weighing 180 kilos (nearly 400 pounds), as well as railway wagonettes and segments of portable buildings. Under the best of conditions, workers would carry these on rigs designed for eight men, four in the front and four in the back. As the report put it, with a vivid simile: under these "crushing and unmanageable burdens," the porters struggled at every step, "resembling in their impotent agitation ants rushing around a prey disproportionate to their size." When paths became steep and slick, it was impossible for eight men to work together, sometimes leaving two to bear the weight alone.

In many areas of the Mayombe, steep, narrow, and slick was the rule, not the exception. During the rainy season, porters were made to carry the hefty cement barrels down riverbanks (sometimes six meters in height), across flooded regions, and through narrow and slippery passages, all of which rendered porterage, in the words of one report, "excessively painful and dangerous." Falling while carrying such a burden presented other hazards as well. "If one of the porters stumbles and falls, he is hit by the handle he's carrying. With the others remaining standing, the one who falls is hit violently by the rig. The falling drum can also hit another porter. In slick terrain this kind of porterage is not possible; new trails are always slippery after a storm." Such heavy loads were exhausting to move a matter of feet; as it was, they were transported for hours on end, mile after mile.

In addition to the purely physical exertion required, porters often worked directly under the Batignolles, meaning that their living and working conditions were beyond the purview of the Labor Service.

Company porters were not provided housing, since they often followed foot trails some distance from construction sites, forcing them to spend nights without beds, blankets, or shelter from rain and cold. Hot meals were never on offer. Rations were distributed at the start of a journey. Hungry men often binged on their food quickly, only to find they would later pass days without eating. Many porters, according to one doctor, developed on their bodies "phagedenic sores"—eroding ulcers—of "fairly large dimensions." Adding to the hardship, *surveillants* often chose the weakest men to be porters. The rationale was that stronger, healthier bodies were needed to clear, terrace, and lay rail in order to keep construction on schedule.

In September 1927, during one of Antonetti's extended leaves in France, the acting governor-general, Joseph-François Reste, took the opportunity of his newfound position to write a pointed letter to the director of the Batignolles about the use of porters for particularly arduous work. He cited an incident in which a doctor told a company agent that a group of weakened workers were not healthy enough to work as porters. The response of the agent was blunt: "Makes no difference to me, I will use them for porterage and they can all go to the hospital . . . or die en route."

It was not the first time that doctors had received such replies from agents. On another occasion, when a doctor had reprimanded an agent for using unhealthy workers for particularly tough jobs, the agent told him, "I don't give a d—, so I am not a doctor who knows how to choose men." Reste thought such an attitude led to even more deplorable incidents, including an episode in which men were made to carry a single 300-kilo (over 600-pound) load by hand for a number of miles along the line when a rail car could have been used. "It will not escape you that such procedures are absolutely reprehensible and inhumane," Reste wrote. He demanded that the company—for its own good, and that of the colony—remedy the "grave situation" without delay.

Far from surveillance, *miliciens* and *capitas* committed what one report called "indiscretions" to motivate their men to move quickly. Albert Londres witnessed one such scenario with some Sara porters. When they dropped their cement, he wrote, "a *capita* slapped them." They picked it up and walked a few feet before dropping it again. "From slap to slap, the cement made it to Kilometer 80." Though generally prized for their stature, these Saras had been transformed by porterage. "Their desolate state seemed to me without name," Londres wrote. "They crawled along the way like melancholy ghosts. The shouting, the slaps did not revive them. One could believe, dreaming of their distant Ubangi, they are groping for the entrance to a cemetery."

Prolonged, intense physical toil; exposure to the elements night and day; yelling and beatings: such was the plight of the porter and the reasons why porterage was the deadliest of jobs. But in colonial records, a number of factors obscured the deaths of porters. No reliable mortality records were kept before 1925. After that date, deaths of men who acted as porters were often not recorded because they occurred outside the camps, while mortality and morbidity records were usually collected inside. As a result, many porters who died were listed as simply missing or as deserters, if at all. Porters also fell between administrative categories in a number of ways: some worked directly for the Batignolles, others for the Labor Service. Still others were men and women who lived in the region who fulfilled their *corvée*, the labor tax required of all subjects, and were not even Congo-Océan workers.

Even with such incomplete accounting, mortality levels among porters were very high. One enterprising young administrator estimated in 1926 that a third of all porters working under the Labor Service died on the job, probably from exhaustion, malnourishment, disease, or a combination. The mortality rate jumped to 83 percent among porters who worked directly for the Batignolles. Putting aside the reliability of these estimates for the time being, the misery of porterage was undoubtedly a well-accepted fact. Workers knew the fate of porters only too well, as many deserted rather than be forced to meet it. Antonetti called porterage the single job on the Congo-Océan "that provides the largest contingent of sick and dead men."

It was a truth known in the halls of government in Paris, as well. In 1927 Henry Fontanier, a deputy in the National Assembly, stated, "All those with some knowledge of colonial issues know perfectly that one of the essential causes of the decimation of populations is porterage." With both concessionary companies and the Congo-Océan in mind, he continued, "Porterage is the cause of death of I don't know how many thousands, and maybe hundreds of thousands, of blacks in Equatorial Africa." Two years later, the devastating consequences of porterage were still a topic of discussion. After Marius Moutet, a socialist deputy who would eventually become the colonial minister, gave an impassioned speech on the high mortality rates among workers on the railroad, a deputy asked for the causes. Moutet did not miss a beat: "Above all, porterage."

———

JUST AS THE EXACT number of porters who died in the fulfillment of their work is unknowable, lasting injuries on the job are very difficult to quantify. Without psychiatric assessment, it is impossible to know the emotional and mental toll of working on the railroad. Living with deprivation, far from home, with hard labor and frequent abuse: workers were prime candidates for mental illness both during the Congo-Océan construction and in life afterward. Officials dismissed depression and malaise among the workers, attributing it to the African's laziness and natural despondency. Paying greater attention to the psychiatric needs of workers would have done nothing but solidify European prejudices about their inferiority. But that many of the thousands of workers struggled with emotional pain and depression seems impossible to discount.

The story wasn't substantially different for physical injuries that resulted in maiming, invalidism, or death. Government statistics that recorded the number of men unable to work did not usually provide details on the nature of illness or injury. Considering the region's climate and topography, as well as the lack of safety gear like helmets, harnesses, and boots, injuries were likely commonplace. In 1925 a medical inspector estimated that the annual morbidity rate in Brazzaville was about 16 percent, with most hospitalizations resulting from physical ailments and injuries. Considering that the Brazzaville end of the line was far safer than most sections, injuries likely befell a notable percentage of the workforce.

And indeed, at least some workers perceived that workplace injuries took their toll. In interviews conducted in Chad in 1974, one researcher found that railroad workers believed more men died from work-related accidents than from diseases. According to workers' memories, landslides, dynamite explosions, and tools caused many casualties. One worker even said he had been assigned to a makeshift morgue where bodies came in daily from the sites. Other personal accounts echoed these concerns. They spoke of falls, of being crushed in cave-ins and slides, of burns and explosions.

Like many aspects of the Congo-Océan, work-related injuries were more explicitly discussed by critics than by official reports. In the early 1930s, Raymond Susset, an aspiring left-leaning politician, traveled across central Africa and penned a stinging account of the administrative shortcomings of the Congo-Océan—"a stillborn railroad." Among other issues, he condemned the contempt with which workers who had been injured or killed on the job were treated. Instead of observing the rigorous labor laws of the metropole, Equatorial Africa inserted clauses into workers' contracts that dated from the First World War. Under colonial law, workers or next of kin could expect to receive a onetime 400-to-600-franc indemnity for being seriously injured or killed. "The result of this contempt for the value of an individual is that I saw at the M'Vouti hospital two magnificent Saras, each having had a leg amputated after accidents in the Bamba tunnel," Susset wrote, referring to the 1,700-meter tunnel in the Mayombe. For their legs, each of the men was paid the "derisory sum" of 400 francs—about half a month's pay for a white employee on the line.

Susset acknowledged that the cost of living for Africans was low, "but it must be said that it's shameful to see a human existence valued at 600 francs" and "a strong and healthy worker paid 400 francs" for an amputated leg. He wondered why the administration did not instead pay the injured a monthly allowance of twenty or thirty francs per month for life. Rather than being a burden on their communities, men remunerated monthly could live "peaceful and honored" in their villages and even be a source of revenue. Certainly reasonable, Susset nonetheless failed to understand cause and effect on the Congo-Océan: shame, whether in death or life, shaped few policies in Equatorial Africa.

# CHAPTER 6

# TROPIC OF CRUELTY

"Here," she said, "in this place, we flesh; flesh that weeps,
laughs; flesh that dances on bare feet in grass. Love it. Love
it hard. Yonder they do not love your flesh. They despise
it. They don't love your eyes; they'd just as soon pick em
out. No more do they love the skin on your back. Yonder
they flay it. And O my people they do not love your hands.
Those they only use, tie, bind, chop off and leave empty.
Love your hands! Love them."

TONI MORRISON, *Beloved*

THE CONSTRUCTION OF THE Congo-Océan was an act of extraordinary violence. While scholars debate the exact meaning and varieties
of violence, it is impossible to doubt the physical pain, psychological
harm, heartbreak, social dislocation, rampant injustice, and trauma
unleashed by the thirteen years of effort in Equatorial Africa. Village communities were torn asunder, families were broken apart,
homes burned and overrun, recruits were tied to one another at
the neck. Men were forced against their will to work in unhealthy
and grim conditions. Women were taken along as wives, acting as

cooks, cleaners, and at times sexual commodities. Every major stage of the Congo-Océan, from recruitment and transport to porterage and construction, was accompanied by acts of physical violence.

Not every European overseer or African *capita* or *milicien* was pitiless. Some were known for their commitment to their workers' well-being. But a variety of accounts of the Congo-Océan are full of incidents of brutality and humiliation. Though few officials were likely familiar with the sociologist Max Weber, the Batignolles and the colonial state certainly claimed a "monopoly of the legitimate use of physical force" on the railroad. Force, in its many forms, existed at every turn: there were laws that compelled labor, armed guards at camps and construction sites, *capitas* with whips, *tâcherons* with sticks and dogs. Even when these tools were not employed, they reminded men and women of the potentiality of force, the promise of brutality. But cruelty—inflicted intentionally and without serving purpose—was also common. Cruelty is, as the political theorist Judith Shklar put it, "the willful inflicting of physical pain on a weaker being in order to cause anguish or fear." Cruelty on the railroad was violence inflicted on men and women that was punitive without cause, retributive without justice; acts of violence perpetrated for the sake of perpetration.

The challenge, then, is how to measure the regularity and severity of acts of cruelty. Accounts of daily life on the construction site suggest that certain kinds of behavior—what one official deemed "relatively light" physical abuse, which included yelling, punching, or beating—were commonplace. Travelers and writers found whipping, kicking, and verbal assault to be standard components of work. At one spot on the site, for example, Albert Londres encountered a number of Saras pounding at rocks. "Some *capitas* conveyed idiotic orders with furor, commanding to push forward and stand still at the same time, to go up and to come down, all punctuated with the usual, 'Go, Saras, go!'" Londres wrote that the men looked at him "with the eyes of suffering dogs as if I brought oil to salve the burns on their backs." The shouting and the hounding, often coupled with physical abuse, were simply part of the workday on the Congo-Océan.

While Londres and others offered vivid descriptions, their visits were short, and many defenders of the railroad rejected their accounts as uninformed or overblown. Gaston Muraz, a doctor who met Londres on his travels in the Congo, dismissed the journalist's account as "literature, literature!" But travelers' accounts were repeatedly corroborated by the administration's own internal correspondence, including reports that often provided even greater detail than did journalists. Still, even with plentiful corroborating official reports, many instances of violence no doubt went undocumented. The most quotidian forms of brutality are arguably the most difficult to quantify. These acts, which became part of the fabric of workers' lives, were so prevalent, so commonplace, as to go unmentioned, unreported, and unpunished.

What stood out as noteworthy acts of brutality—that is, incidents demanding administrative investigation—must be understood within the context of Equatorial Africa. Long before the first shovel of dirt inaugurated the construction of the Congo-Océan, the entire region had become notorious for its brutality, both for the alleged "savagery" of certain African groups and for European atrocities. Much of this notoriety stemmed from stories of the transatlantic slave trade, of combat between African communities, and of the murder of early European missionaries, merchants, and officials.

By the 1920s, colonization had witnessed frequent armed conflict between French forces and recalcitrant local populations, in turn creating social strife and clashes among African societies. Many Frenchmen liked to think that colonial administration brought order and control to the region, but it certainly did not. Shortages of resources and manpower meant that the French vied for power in

ways very similar to how ethnic groups had established influence for
at least a century: through alliances, trade, taxation, intimidation,
and war. Colonial politics relied on conflict that ran the gamut from
bullying to brutality.

European belief in the violence of Africans, however, was inspired
as much by racism as by reality. Europeans, including French men
and women, had long stereotyped indigenous populations with a
litany of racist—and often contradictory—characteristics. Depending on who was providing the description, Europeans portrayed
Equatorial Africans as everything from bloodthirsty, irrational, and
cannibalistic to weak, docile, and lazy. While Europeans certainly
died at the hands of indigenous populations in Equatorial Africa,
the alleged violence of the African was outstripped in quantity and
severity by documented cases of white brutality, as the horrors of the
concessionary companies made plain. The investigations of Brazza's
commission failed to make incidents of cruelty and killing disappear.

What emerged after decades of conflict was a tacit acceptance,
on the part of an often ambivalent administration, of a wide range
of violence. Before the First World War, for example, the Kouango
concessionary company tried to overthrow an "intelligent native"
leader, named Barambaké, causing a war that resulted in the death
of "numerous natives and several whites." The company was entirely
responsible for the unfortunate events, but the ministry chose not
to punish it. Such a laissez-faire approach was typical in Paris. A
different concession was found to be responsible for "habitual brutality toward the natives." The ministry went only so far as to ask
the company to stop. Accountability was a fate few white men
ever faced. Even when inspectors found business interests guilty of
causing deaths, the ministry in Paris—not to mention the colonial
administration—did little to punish the companies.

Not all abuses were perpetrated by concessions. Colonial officers
were implicated in sometimes gruesome acts of cruelty, as well. In

1913 a French captain named Teulière was accused of the "mistreatment" of over two hundred "Arab" men, women, and children near Bangassou in eastern Ubangi-Shari. What qualified as "mistreatment"? Suspecting them of trading slaves and munitions, Teulière and his men surrounded their village before dawn, captured them, and imprisoned them—a "purge" that was carried out "with particular brutality." In the process, Teulière personally demonstrated to his men how to whip prisoners with the *chicotte*. He then acted as prosecutor, judge, and jury, condemning more than thirty of them to fines and imprisonment. As a final humiliation, he "distributed" the Arab women to his soldiers—a not-subtle euphemism for rape.

Six of the men Teulière abused provided greater detail of the captain's brutality. They traveled to Bangui, as they put it, to "demand justice from the chiefs of the French" and to recount their abuse. "Rachid had an eye punctured," they testified. "Abderamam had an eye punctured." Other men had their money, goods, and wives taken from them. The men explained they had had no reason to expect such treatment; they'd paid their taxes when asked. And yet "everything we had was stolen." In all, they lost 35,000 francs, four thousand kilos of rubber, merchandise, and chickens, and "our huts were burned or torn down."

This was not the only example of Teulière's *phobie* of the Arab population. On multiple occasions, and without being ordered, the captain physically harmed and tried to expel Arabs, many of whom had lived in Ubangi-Shari for fifteen years. He burned villages both when the local chief refused to meet him and when villagers, terrified, fled to the bush. On one occasion, he burned forty hectares of crops that were nearly ready for harvest. His superior condemned such acts as "inadmissible" and "useless and maladroit"; they were "proof of the greatest inhumanity." But the captain remained unrepentant, mincing no words about the Arab population he suspected of slave trading. Their proximity to his post was, to his mind, "an insult to our flag." His solution was to "attack them" with one hundred African soldiers armed with rapid-firing rifles—a decision he felt fully justified in taking. "I will not make them flee," he wrote to his superiors, "I want to exterminate them."

Such rogue violence was not limited to Teulière. And as objectionable as he might have been, his behavior was at least based on some—if a deeply misguided and repugnant—rationale. Other officers and administrators tortured for little reason at all. In 1915, for example, an official in Middle Congo was accused of crushing a man's testicle with a hammer while trying to extract a confession to stealing five hundred francs. Another administrator allegedly smashed a chief's hand in a printing press before having him whipped; his wounds were then covered in honey and the chief was bound to the ground to attract bees. One French colonel even reportedly asked his soldiers to string a necklace of the ears of those they killed. There is little evidence that these crimes were seriously investigated. Antonetti claimed they were, but he also insisted that brutality was more common in France than in the Congo.

Violence remained a part of life in the interwar years, when accounts of European cruelty appeared more frequently in books, travel journals, and the press. White brutality represented a problem for the colonial state not because—or at least not primarily because—of the harm it caused. Rather, shocking incidents threatened to draw unwanted criticism of colonial policies, including the building of the railway. In his travel writing, Robert Poulaine recounted meeting an experienced magistrate known for his good professional conscience who confirmed the brutality the colony was known for. "What worries me the most in this country," he told Poulaine, "is the mentality of the white vis-à-vis the native." The magistrate said he had never once seen a report of a crime committed by an African against a white man. "By contrast," he continued, the files were full of dossiers about whites who committed "brutalities, extortions, abuse . . . the victims of which were natives." Some of the crimes were too awful to describe. "Modesty and national pride" inclined Poulaine to censor details in his own book.

Other writers were not so impeded by pride. In his 1934 series in *Le Populaire*, Pierre Contet detailed the quotidian brutality that infused many aspects of Africans' interaction with Europeans. His series seamlessly linked what he called the "*bagne* [prison] of the Congo-Océan" with the systemic violence of the colony as a whole. Contet's subject matter ran the gamut from the recruitment of "volunteers" to the impact of health policies, from famine and food shortages to imprisonment and taxation. His details, as well as his references to specific cases of mistreatment, made for unrelentingly dark reading. For months, regular articles provided insight into the brutality of colonial life for France's equatorial subjects.

Contet's assessment of the region ran from what is often called "social suffering" or structural violence—the misery and hardship that stem from systemic iniquities within a society—to the most intimate and lurid forms of interpersonal violence. For example, Contet focused on the deep implications of taxation, in the form of both monetary taxes and labor corvée, on the population. In some cases, local companies imposed their own taxes, without the use of accurate censuses. But failure to pay the charges levied could result in imprisonment and the "legal rape" of debtors. All colonial subjects—men and women of all ages—were required to pay the annual labor tax; it was, in theory, five days of work per year, usually tending to roads or other public works projects. But the corvée was abused with impunity, often being extended to periods of weeks or months, during which workers were neither paid nor fed. Without friends or family to help, he said, workers would "croak on the road, like dogs."

From such simple policies, more disturbing consequences stemmed. Sexual violence—a taboo subject that most commentators avoided or mentioned only in passing—was a theme Contet returned to more than once. He documented its myriad, often sadistic forms in the colony. Girls and women were regularly excused from their corvée duties, he reported, only to be housed in the huts of African *miliciens*. There, the women "are copiously fed, but they pay for the favor with their bodies by will or by force." When monetary taxes were demanded, impoverished husbands and fathers sent their "young and pretty girls" onto the streets where "single white men . . . for an hour, rented their charms." In colonial prisons, inmates were stripped naked, and the *chicotte* reigned supreme; "girls and women are raped, every night," and all learn that "the white man or his servants (blacks of other countries) are their most terrible enemies."

Should Contet's readers have thought that only representatives of the state were driven to such depravity, he described in detail the experiences of an African boy in Gabon who was accused of stealing—ironically—a *chicotte*, one of the hide whips used to punish employees. The story broke first in 1932 and was picked up in many colonial publications, even generally pro-colonial ones. The boy, named Massima, was tortured by his white employers into confessing, first by being tied to a tractor and dragged. When this spectacle terrified his fellow workers into fleeing in to the bush, his torturers stopped briefly, but apparently only long enough to intensify the child's misery. They then tied him to a Saint Andrew's cross; a white woman, driven by "a shameless sadism," encouraged her friends to burn his genitals with a blowtorch.

When their tortures failed to elicit an admission of guilt, the Europeans decided to waterboard him by hanging their suspect by his feet, his head partially submerged in a barrel of water beneath him. Making an occasion out of it, the party drank champagne and made bets as to how long it would take for the boy to drown. When he died after five minutes, one of the white men was declared the winner. They then drank their aperitif in the presence of Massima's still-hanging body. His killers tried to hide Massima's body, burying it in a hole that they ordered their servants to dig. But Massima's brother watched them from the bushes; once they left the scene, he partially exhumed his brother and cut off his hand to show to the local commandant to prove the crime.

The handling of Massima's murderers reveals the extent to which white men and women in the colony felt unfazed by their own brutality. Once the crime was reported to the authorities, the murderers openly admitted their guilt. The judicial process that followed, Contet argued, was conducted by individuals who knew nothing of the law but were acquaintances of the accused. The cruelty of this crime was exceptional, he admitted, but the judgment against the accused was not: they were all acquitted.

From punching to raping, from burning villages to torturing and murdering, white abuse of Africans was a regular part of life in Equatorial Africa. This context profoundly shaped the way colonial officials acknowledged, understood, and investigated the violent mistreatment of workers on the Congo-Océan. As a result, more quotidian abuses were largely left underrepresented or undiscussed in the voluminous paperwork generated by the construction. Even the director of the Labor Service on the Congo-Océan, Jean Marchand, spoke in casual terms about the regular mistreatment of workers. He spoke of African workers being treated "too roughly," and of "hits given"—observations carefully made in the passive voice to avoid drawing attention to the perpetrators. It bears noting that in the 1920s and '30s in France, very few legitimate employers physically abused the workforce. Certainly, abusive behaviors, especially sexual harassment, were a part of French employees' experiences. But episodes of physical abuse—such as foremen punching workers—were rare and were met with absolute intolerance, leading to walkouts, strikes, riots, and criminal investigations.

Colonial archives and personal accounts, then, provide a dis-

turbing picture of the relative frequency and severity of violence—
but it is, even then, certainly an incomplete one. Acts of abuse were
investigated only when public opinion, racial and political stability,
or productivity made it untenable for the administration to ignore
them. Marchand admitted that many incidents were simply overlooked in the interest of completing the railroad quickly. The Labor
Service "closed its eyes to certain means of too expressive persuasion" used by agents of the Batignolles. These methods, he reported,
"chided" the men and got them back to the construction site, "even
when their skin carried traces of some ecchymosis [bruising]." He
insisted that when the day came when a worker was "seriously man-
handled" and one could "clearly establish the guilt of an agent," then
the guilty would be prosecuted to the fullest extent of the law. Such
clarity, however, was very rare indeed on the Congo-Océan.

While very few white overseers were ever seriously punished,
their deeds were nonetheless often the subjects of detailed investigation. These investigations provide insight both into the quotidian nature of beatings on the line and into the range of brutality
that workers came to expect while working. Far from portraying
the recipients of abuse as passive victims, investigations into the
mistreatment of labor represent, in fact, one of the rare opportunities where workers expressed, on the record, their experiences of
the railway. If justice was rare, workers nonetheless voiced moral
reproach of nearly every aspect of the project.

———

IN 1925 TWO *CAPITAS* near Mavouadi reported that one of their
workers had been beaten with a stick by a white overseer. The worker
died of contusions a few hours later. As a result, the entire team of
workers fled into the forest. The murder of this worker cast in relief
how complicated it was for the colonial state to deal with European
brutality on the railroad. The Labor Service, which was still new and

untested in 1925, first and foremost wanted to avoid a scandal. But, beyond that, it was never easy to know which side to take.

One way to reassure workers was by punishing abusive overseers. But the Labor Service faced regular complaints from the Batignolles that investigations into acts of cruelty undermined Europeans' ability to command respect and discipline from their African subordinates. Many white overseers claimed the only way to motivate their workers was with punishment. When threats escalated to violence, they ran the risk of having their workers refuse to work or bolt into the forest. The delays involved were costly to the company and reflected badly on the foremen, who had daily tasks to fulfill. It also meant more work for the administration, which had to find replacement workers to meet the Batignolles' needs. Punishing overseers' abuses, however, was tricky as well. The 1925 episode, for example, allegedly caused "an exaggerated emotion" among the European personnel of the Batignolles, who resented being openly criticized for their poor judgment. "Now that one of the overseers has been humiliated by a confrontation with the 'nègres,'" Marchand explained, "authority is no longer possible."

The belief that whites should never be held accountable for physical brutality was a long-standing assumption in Equatorial Africa (and in many other of Europe's colonies), where racial boundaries and hierarchies were carefully guarded. There was an assumed "practical" reason for this unwritten policy. That workers needed to be verbally or physically motivated to work was a mantra on the railroad. Any erosion of the white man's right to treat his workers as he saw fit allegedly posed grave dangers. Access to violence was necessary for upholding white authority.

Who were the white men who took up such jobs? Soon after construction began, the Batignolles found it increasingly difficult to find Frenchmen willing to work on the construction site. Engineers and high-level administrators in Pointe-Noire remained overwhelmingly French, but the *surveillants* and, especially, the *tâcherons* were not. The company resorted to hiring whomever it could, often regardless of work experience, technical skills, or moral conduct. The new employees became a constant source of concern for French officials who were wary of having white men of dubious back- grounds and temperaments in close contact, as well as conflict, with workers. Colonial authorities quickly grew nostalgic for the early French employees, who were wistfully imagined to be well chosen and of upstanding morals. For its part, making the most of the work- ers it had, the Batignolles doubled down on defending the propriety of all its white workers.

White employees constituted a ragtag and ever-changing group of men. In 1925, on just one twelve-mile length of the line, the *sur- veillants* and *tâcherons* included a Briton, four Italians, a Portuguese, a Belgian, and four Greek brothers. The director of the Labor Service at the time, Georges Thomann, doubted that these "adventurers com- ing from the four corners of the globe" had their workers' best inter- ests at heart. He had good reason for concern. Three of the men spoke hardly any French. The Portuguese, who had the important role of *surveillant,* was only twenty-two years old. And while the Briton was fifty-one, he was a former hunter who had apparently been kicked out of the Belgian Congo for undisclosed reasons, spent years on the move, and had a reputation of being "extremely hard on the natives." Should their biographies not be worrisome enough, Thomann did not mince words in his summation: "on the whole, these men are parasites, for whom the only goal is to rip as much as possible from the country without concern for what they leave behind."

From its inception, the Labor Service was displeased with the Batignolles' failure to discipline their white employees, especially the non-French ones. By 1925, the white employees being hired on to the project were entirely "unqualified to establish authority" over their men, resorting to teasing, bullying, or worse to get their workers moving. They didn't let their workers circulate during the break in the middle of the day. And if the overseers were displeased with them, they refused to allow them to commute in wagonets, a much less tiring means of transport to and from the site. While such poor management skills deserved some attention, they failed to capture the real potential for violence that existed in relationships between some white men and their workers. Indeed, some of these "adventurers" hired by the Batignolles proved to be brutal and even murderous.

While it is difficult to have sympathy for abusive men who often showed themselves callous to the miseries of their men, the life of the European foremen was far from easy. Overseers often lacked gusto for their jobs. They were generally poorly paid for "painful work"; incomes ranged between 800 and 1,200 francs per month for *surveillants* to upward of 3,000 for men with more technical knowledge (as compared to the fifty or so francs a month their workers earned). The salary was similar to what they would have made in Europe, but their living conditions were "particularly hard." Indeed, unlike in Europe, a high percentage of white men on the Congo-Océan became seriously ill with fevers or intestinal ailments; and dozens died from tropical maladies contracted during their tenure.

For those men who survived, loneliness was a constant companion, as many seldom encountered other Europeans, going weeks or months at a time without conversation or news of the outside world. Supplies of fresh food were infrequent, meaning they often lacked bread and wine, staples of European diets. Unlike at home, where after work they could relax or go to the cantine, in the Congo, evenings were spent, as one rather poetic report put it, "crushed under the vault of the virgin forest, harrassed by insect bites, surrounded by boys they don't know how to direct and with whom they often can't communicate." One French engineer who'd lived in the United States, Australia, Persia, and Tibet, from low deserts to the highest mountains, said he'd never faced such difficulties as he did in the Mayombe. "Can anyone blame them," one Labor Service report asked, "for a lack of enthusiasm?"

For the men who worked the construction sites, the physical and moral tribulations of their white overseers offered cold comfort when violence occurred. "Mr. Bowen is always hitting us, even for no reason," a *capita* named Mounguiti reported casually in 1925. But on some occasions, Bowen's aggression caused real damage. "This morning he kicked me during roll call, a kick that didn't hurt much," Mounguiti continued. "On the construction site, he socked me with violent punches in the sides and kicked me six times in front. At the same time his dog jumped on me and tore my loincloth." The beating did not seem too harmful at first. "At the moment, I didn't feel much at all, then, all of a sudden, the pain became severe, I started trembling and fell. My comrades cared for me and took me to camp. Now my lower stomach still hurts." Mounguiti ultimately was taken to the hospital. A sign of the extent to which some workers internalized the hierarchy, Mounguiti's workmates said he kept apologizing as he was being beaten, crying "Pardon, Monsieur!" The white man kept hitting him.

For his part, Bowen, the fifty-one-year old Englishman who had apparently been kicked out of the Belgian Congo, showed no remorse. He admitted in poor French to the Labor Service that he'd hit Mounguiti but said he did so only once. Then he explained, "The men don't want to work, it takes them an interminably long time to do almost nothing, I am obliged to shake them a little." Uncommon in such instances, the investigator, named Galoisy, pushed deeper. He visited Bowen's camp and found that a number of the workers bore the scars of the *tâcheron*'s punishments. One worker, Bako, had been whipped twenty-nine times with the *chicotte* and had scars on his shoulders. Two other men, Madingou and Massouna, had faced similar treatment. Moussa had been kicked in the face and still had contusions. Massika had an injured hand from being hit with a stick.

Bowen beat and broke the thumb of a man named Makouma while he was carrying the European in a sedan chair to Pointe-Noire. Bowen repeated claims about the "extreme laziness" of Africans. As the Labor Service itself accepted that some workers needed pushing, Bowen's position remained impregnable; he was let off with a warning.

Bowen's aggression was but one sign of the pervasive regime of brutality on worksites. The same month he beat Mounguiti, a twenty-three-year-old Portuguese *tâcheron,* Perrera, also injured a worker. He had found his men not doing their job of clearing a stretch of forest along the rail line. He picked up a piece of wood and threw it at one of his workers, Tombe, but missed. He then took up a larger piece, about the size of a baton, and threw it again at Tombe, some forty feet away, striking him in the head. Dazed and bleeding, Tombe was led to a medical post by his workmates. Perrera's men also alerted the neighboring work teams, five of which—168 men in all—walked off the job in protest. Perrera claimed he did not mean to hurt Tombe. It seems that Perrera was not punished in any way; his excuse, that he was motivating lazy workers, was indisputable on the Congo-Océan.

Who could investigators trust, and what could they do? The questions plagued them for years. The continued ambivalence helped establish mistreatment as the norm. From the outset, the racial hierarchy of the construction effort meant that Africans were not seen as reliable witnesses of their own abuse. White supervisors insisted that workers would deceivingly claim mistreatment to get out of work. Not all officials agreed. Galoisy, who investigated Bowen's abuses, believed that workers had no reason to accuse someone who was not guilty. But his concern also did not change overseers' behavior. Workers learned quickly that making complaints was more likely to anger their bosses than to result in any form of penalty, let alone in justice. Complaining had little chance of improving workers' circumstances; officials rarely reprimanded anyone seriously, let alone fired them. Even workers who came forward with visible scars were turned away as lacking evidence.

Angered that rumors of brutality scared off recruits and caused anxiety among workers, the administration insisted that all investigations had to have "precise facts" and not be based on "simple rumors" that lacked merit. But precision was exceedingly difficult to come by, especially when workers' accounts were discounted as biased or intentionally misleading. Even medical reports by doctors were often noncommittal when it came to abuses. For example, in the case of two workers, Issongo Ilinga and Okamba M'Boue, who bore "scars" on their shoulders and back, a medical examination determined that "no medical conclusion [was] possible for the degree of violence" and reached "no conclusion on the origin" of the wounds. One worker who was beaten to death by his white overseer was even said to have had "no exterior wounds." With neither witnesses nor science to legitimize complaints, beatings went unpunished.

Only in the face of blatant evidence was punishment likely. In 1925, for example, a European named Moutini beat a worker named Maboukou with a machete, causing him wounds that left him unable to work for thirty-seven days. Moutini was ordered to pay a fee of 150 francs, though he later appealed the decision. At the same time, another white man, Paoli, was found guilty of murdering a man named Makosso and was punished according to the law. He served the entire sentence handed down: six months in prison.

The lack of consequences for white employees meant that violence became normalized. In 1924, for example, multiple men working for Nicolas Caraslanis complained of being slapped, while one of them, Massengo, bore scars from a baton. Caraslanis also kept his workers at the construction site until well after quitting time, denied them the regulated time for lunch and afternoon breaks, and docked their pay according to his whim. An investigation launched against the European found insufficient evidence. The administration reminded him of the need to follow regulations regarding work hours, payment, and abuse. But a year later, Caraslanis's workers were still complaining.

In May 1925, his team abandoned their work again to register complaints of mistreatment. Some of them showed scars, the origin of which could not be established. Others, however, showed "obvious marks from violent hits from a *chicotte* or cane." At this point, Thomann of the Labor Service expressed with greater urgency to the Batignolles the need to end the abuse, insisting that he would pursue legal action against Caraslanis if the mistreatment continued. But in a sign of the importance of keeping up appearances, Thomann also found fault with the abused workers. To preserve "the authority" of the white foremen, he reproached the workers for leaving the site to make their complaint. Were they to do it again, he warned them, they would be "severely punished." It is unclear what happened after that. Perhaps Caraslanis changed his behavior. It is also very possible his men were too scared to complain again about abuse. For the administration, either result was an improvement.

⸻

WHILE THE LABOR SERVICE had investigated mistreatment of workers on the construction site since its creation, their interventions became more rigorous around 1927. At the time, the Labor Service was under the direction of a Mr. Titaux, who proved decidedly more aggressive than his predecessors, Marchand and Thomann. Investigations brought conflict with the Batignolles, which resented even sporadic efforts to police the foremen's behavior. The company, from the head engineer to the lowliest white overseer, regarded its African workforce as the most frustrating impediment to progress on the construction site. It repeatedly expressed exasperation that the administration did not allow the company to discipline—that is, to threaten, beat, or imprison—African laborers.

The company considered labor stoppages and desertions—regardless of the cause—to be costly and unnecessary. The governor-general's office, under constant pressure from Paris to get the railroad built, was sympathetic. But the Batignolles was also loath to accept that workers' indiscipline was caused primarily by the brutality of their own European employees. Rouberol responded to inquiries about mistreatment with impatience and frustration, and usually with counterdemands that the colony provide the number of workers promised in their agreement. The competing pressures chilled relations between the Batignolles and Antonetti, but it also started to drive a wedge between the governor-general's office and the Labor Service.

Relations became particularly tense in 1927 when, during one of Antonetti's visits to France, the Labor Service intensified its investigation of work-related violence. A series of incidents between Batignolles employees and their men led to desertions of entire teams. According to the foremen overseeing the teams, the workers had made unacceptable progress on their tasks and, when threatened with punishment, abandoned their stations. The response of the company was swift: a telegraph from company headquarters to Titaux at the Labor Service demanded that he "energetically sanction the teams of workers" for leaving their worksites. Titaux, however, proceeded methodically: he toured the various worksites, interviewing European foremen and engineers, and meeting with laborers. For the African workers, it was a rare opportunity to speak with what seemed like a sympathetic Frenchman. For company officials, it reflected the Labor Service's misguided propensity to take the side of the worker.

Much to the Batignolles' dismay, the Labor Service's investigations showed that workers deserted their posts after suffering ill-treatment. Investigating cases of desertion that occurred in September 1927, Titaux found that work was suspended only after the European overseers had beaten a number of their men. Alfred Antheaume, a French foreman, had told his *capita* to beat men who weren't working hard enough. In a deposition, the *capita* described what happened next:

Q. You quit working yesterday. For what reason?

A. M. Antheaume, determining that my men were not working enough, asked me to beat them with a stick. As I had cut a small branch, he ordered me to cut a stronger stick. I responded to him that it was prohibited to hit. Arguing that no one was working, M. Antheaume beat the men of the team, Gamambi, NGoulou, Mbou, on their sides, using a thick stick. He held the stick in two hands and hit with force. As these brutalities were also committed again on the persons of MoKouba, Mbou, Mokass I, [and] Gama, we went to Mavouadi to tell.

The Batignolles chose to deny it, showing little concern for discovering what had actually happened on the site. Failure to punish African indiscipline, the chief engineer of the company argued, would lead to "anarchy spreading across the entire worksite." Such "acts of disobedience" were the result of workers believing they could "defy with impunity" their foremen, encouraged by "the weakness with which the [Labor Service] receives our complaints" despite being well founded. Further inquiry, however, showed that the Batignolles had been too quick to defend Antheaume. The Frenchmen later *admitted* he had beaten workers—but, he claimed, only two of them, and with a switch, not a stick. The difference between a stick and a switch was likely determined by which end of the beating one was on.

Such incidents were not isolated. Another foreman, Redi, working nearby, was likewise accused of beating two workers about the shoulders with "a big baton," again resulting in the workers leaving. The construction company wanted the workers punished. But rather than reprimand the workers, Titaux believed a different remedy was in order. "Not only can I not prescribe severe punishment for these teams," he wrote to the Batignolles, "but I also have the honor of asking you to take all measures you consider useful to put an end to such incidents."

But the Batignolles remained resolute in defending its white employees, often at the expense of the African men brought to build the railroad. A powerful example was the case of an Italian overseer named Merignani, who had earned a reputation among his workers for "extreme brutality." The Italian beat his workers with the thick handle of a tool, probably a shovel or a pick. In 1927 Titaux reprimanded him, finding it particularly relevant to investigate this incident because another Italian had broken the arm of a worker who was trying to deflect the force of the Italian's blows. Violence disrupted progress on the line, Titaux pointed out, and if the beatings continued, Merignani's two teams would refuse entirely to work for him. The response from the company was anemic at best. A Batignolles engineer refused to accept the veracity of the complaints but did send word directly to Merignani that he "must not brutalize the blacks."

Other company employees were even less responsive, turning the accusations back on Titaux and his efforts to interfere. Defending "this excellent foreman" against accusations of "supposed brutality," section head Bernet argued that Titaux's concern for workers caused more trouble than good. The real culprits were the Africans who *provoked* violence in order to get their foreman in trouble. "Since the arrival of Mr. Titaux," Bernet insisted, "the natives have shown an indolent and provocative attitude toward Merignani *in order to get hit* so they can go make a complaint against him." Bernet ignored Titaux's findings and cleared Merignani of any wrongdoing.

Bernet claimed that the man who had been beaten, along with the witnesses, had lied. Worse still, in Bernet's opinion, was that Titaux hadn't even interviewed the white men on the site.

The Batignolles' indifference to workplace violence, in the case of Merignani, led to tragedy. The investigation did not cool the Italian's aggression. Six months after the Labor Service reprimanded him for beating a man—and after his white co-workers ardently defended him—Merignani beat and stomped to death his *capita,* Garassena. A sign of how quickly a worker's life could change on the railroad, Garassena left camp for the construction site after lunch, only to return three hours later. He reported that "a white Italian" had punched him repeatedly in the chest, kicked him in the stomach and sides when he fell to the ground, then hit him with a baton. He was sent first to a nurse, then to a medical camp, where he died. An autopsy discovered that violent trauma caused internal hemorrhaging, including a burst spleen.

Merignani tried to defend himself by claiming that Garassena was weak and sickly and that for every day Garassena worked, he had spent two days in camp. But others described the *capita* very differently: they called Garassena a "colossus, of the Sara race" who did not miss a single day of work from November to February 1928. Under investigation, Merignani's credibility broke down. He later tried a different tack, claiming that he would never have dared attack Garassena because "he was so strong that with one punch he would have crushed me."

Titaux did not doubt Merignani's culpability. In addition to identifying his inconsistent statements, Titaux repeated the Italian's reputation for callousness. He'd been previously linked to beating not only his own men but two *capitas* in a neighboring team of workers as well. That team had been under the direction of two Frenchmen, Patra and Mary, who confirmed Merignani's abuses. The Frenchmen admitted they hadn't previously condemned him because they dreaded retribution from the Batignolles. In fact, Bernet, the Italian's ardent defender, had written the Frenchmen's statement and forced them to sign. Fearing being "shown the door," they even admitted to supporting the embattled Merignani, "disregarding all principle of humanity towards the natives."

More revealing still was testimony given by a worker familiar with Merignani and his fellow *tâcherons* and *surveillants*. An eighteen-year-old Catholic convert, Paul Mboungou, had received Garassena after he'd been beaten and had sent him to the medical camp when he assessed he was "too sick" to work. Mboungou took roll call in the morning and kept the books on who worked and where. He knew no details of the attack on Garassena, but he provided a sense of the scene on the worksite. Having asked about his encounter with Garassena, his interrogator continued,

> Q. Is that all you know?
> A. Yes that's all I know, but Mr. Merignani hits the men too much for nothing, even those who work. There are, along with Mr. Merignani, other whites. First of all, Mr. Tardif . . . and Messieurs Ledro and Fleury. These latter also hit the men: Mr. Ledro a little, and Mr. Fleury a lot, the same as Mr. Merignani.

These other men weren't under investigation, but Mboungou did not miss his opportunity to speak against them. He did not condemn every white man. He explained that his boss, Pelgas, had him mark as present at work all the men who were injured from abuse; they had the right to get paid, Pelgas had reasoned, even if they were too injured to work. But it is clear that Mboungou saw many violent men among those he worked for.

> Q. Do you have anything to add?
> A. I have nothing to add, except that the whites of the

Batignolles are not happy with me, because I challenge
their employees, and they are always in dispute with me.

If Mboungou felt empowered to speak out in light of Merignani's
obvious guilt, he fell victim to his own youthful optimism. A pre-
liminary judicial inquiry ruled on procedural grounds that the Italian could not be charged in Garassena's death. Although less than a
week passed between the *capita*'s death and the bringing of charges
against Merignani, the general prosecutor ruled that the administration had not moved in the timely fashion required by law. The
medical examiner also refused to state conclusively that Merignani
had caused the *capita*'s death. Merignani was not charged.

The aftermath of the murder of Garassena revealed the ambivalence and conflict at the heart of the administration's surveillance
of violence. By the time Merignani walked free, Governor-General
Antonetti had returned from his leave in France. Aware of the
Batignolles' criticism of the policing of abuses, Antonetti chose to
take the opportunity to reassess the Labor Service's function. The
governor-general told Titaux that the Labor Service was essential to
continued surveillance of the construction site, the camps, and the
behavior of all workers. But he also emphasized the need for coop-
eration; it was "equally indispensable" to maintain "a spirit of harmony" with the directors of the Batignolles. In so stating, Antonetti
effectively shifted blame for Merignani's brutality away from the
company and onto the Labor Service. The incident, he said, was
"clear evidence of a lack of agreement."

Titaux refused to accept his superior's assessment of the situation. But all he could really do was complain about his continued
challenges, working nearly alone, trying to investigate and punish
the frequent cases of brutality on the railroad. After the Labor Service's conflict with the Batignolles in 1927 and 1928, its surveillance
of violence appears to have been less vigorous. The Labor Service's

reputation was no doubt further tarnished just days after Merignani's absolution when one of Titaux's associates, Paul Renel, was accused of beating an African *capita*. Renel did not refute the allegation; he was dismissed for "indiscipline."

Meanwhile, Merignani returned to the worksite, no doubt relieved but perhaps also reassured. While touring the worksite, Albert Londres encountered an Italian overseer who was possibly Merignani. Londres's Italian overseer, who had returned from Pointe-Noire where he had faced an inquiry into the death of a worker, was indignant at the process. He spoke disdainfully of even suggesting he was guilty of killing someone. Such trials, the Italian claimed, were pointless; no one, he insisted, killed Africans. "We don't hit the blacks to kill them but to make them work," he told the journalist. "Kill? Kill? When my hurricane lamp is out of fuel it burns out; if I blow on it, the flame doesn't last. I blow on a black, I don't kill him. So, I was acquitted."

———

THE MEN ON CONGO-OCÉAN continued to complain about acts of brutality, but less and less was done to punish perpetrators or to end the cycle of violence. It is very possible that the Batignolles adopted new strategies for dealing with administrative inspections, including deception. Albert Londres recounted the visit of Dr. Lasnet, a government inspector from the Health Service touring worksites in the Mayombe. In his typical playful style, Londres wrote as if speaking to Lasnet: "Mr. Inspector, you have certainly missed a beautiful spectacle. It was organized specially for you. You come believing the vicious people who claim that the blacks die on the construction sites of the 'Batignolles'? We're going to show you how they are treated."

Londres claimed that the site's regular workers had been replaced with "model detachments" brought in for the inspector's benefit. The sickly laborers on the site were hidden in the forest, out of sight.

In their stead, Potemkin recruits were "handsome men" chosen in Brazzaville and well supplied with khakis, knapsacks, blankets, cutlery, tea, a towel, and soap. They were brought to the worksite to meet the inspector in covered trains, six to a car, and fed pâté and sardines. "All they lacked was a hot water bottle under their feet and a good cigar!" They even had a warm meal, Londres noted, prepared under the careful watch of a white woman. There was no doubt some hyperbole to Londres's account; but preparing for inspections was certainly more effective than responding to critiques.

By the end of the 1920s, much of the work shifted increasingly to the eastern end of the railroad, where the administration had direct control over the construction effort. While government officials did not have to worry about interference from the Batignolles in cases of abuse, they still remained tepid in their policing of brutality. The abuses they alluded to were no less serious. Beatings remained a common theme: the Greek employee Caraslanis, for example, broke the leg of a worker named Tchimbana by hitting him with a wooden plank. And the excuses and extenuating circumstances changed little: Caraslanis denied having hit his employee with a plank, saying he had only punched him; a medical examination said that Tchimbana had only light swelling on his cheek and that his fracture could have been caused by a fall rather than a hit. The fact that Caraslanis had previously been accused of mistreating his workers for some five years still had little impact. There is no evidence any action was taken against him.

In the final years of construction, workers continued to be menaced by European overseers driven by illness, addiction, and sociopathy. One white *tâcheron,* for example, raped a male African worker, tried to kill the son of a French official, and then committed suicide—an episode that, for reasons that remain unclear, seems to have been nearly expunged from the files. Sexual violence was rarely discussed in official documentation beyond fights that erupted between workers over women. Prostitution, same-sex liaisons, and rape were undoubtedly a part of life on the railroad. The fact that this case—which included attempted murder and suicide—appears in only passing reference in the archives suggests that officials simply turned a blind eye to sexual crimes or kept evidence of them in separate files.

Other vices were more readily used to explain, and even excuse, brutality. In 1930 a military officer named Roustan, stinking of alcohol, violently beat "without any motive" a fifteen-year-old recruit who had allegedly looked askance at him. When a sergeant intervened, Roustan tried to strangle him and continued to hit the boy. A doctor and ten soldiers had to get involved before Roustan, who was said to be suffering from delirium tremens, could be controlled. Roustan was condemned not for the harm he inflicted on a fifteen-year-old but for the scene he made in front of Africans. He later admitted it was "unacceptable for a white man" to show such a lack of "dignity in front of the natives." Roustan's embarrassment as a white man—rather than remorse for his violence—was shared by the governor-general, who acted swiftly to try to quell a scandal.

Throughout these episodes, colonial officials continued to ponder what level of violence "deserved" investigation by legal procedure. A victim's hospitalization in a medical camp was a proposed standard for triggering a judicial inquiry. It was put to the test in January 1933, when three Europeans were scrutinized for having "brutalized" workers to the point that they needed medical attention. One worker, N'Gamma, spent two days in hospital with internal pains; another, Paye, was beaten by three Europeans but not hospitalized; and a third, Tebro, was beaten by the head foreman, Do Rio, and missed ten days of work and bore bruises on his head. Despite the apparent seriousness of the injuries, a judicial investigation was not ordered. Instead, the incidents were said to have lacked gravity and the "climate and the ongoing fatigue that the Europeans suffer" were deemed mitigating factors. The head of the Judicial Service did not contravene, echoing the administration's lack of concern; none of the cases deserved being called an "affair."

The cause of N'Gamma, Paye, and Tebro was unlikely to become a sensation in any newspaper, in the perverse measurement of justice on the Congo-Océan. Pain or fatigue, questions of right and wrong and fairness: such concepts were extravagances on the railroad. Until its completion, few things guided French policy regarding the physical abuse of workers more than potential embarrassment. For the men and women on the wrong end of cruelty, the only shards of hope could be found by taking matters into their own hands.

# CHAPTER 7

––––––––––––

# DISOBEDIENCE
# AND DESERTION

As DEHUMANIZING AS MANY aspects of the Congo-Océan were, men and women had myriad ways of expressing their humanity, most of which were probably lost on their white overseers. If the language barrier on the construction site confused communication among workers, guards, *capitas,* and white employees, it also enabled Africans to discuss, criticize, or even mock the Europeans they disliked, no doubt occasionally right in front of their faces. The subtle facial expressions, the bodily gestures, the clipped utterances that are a part of all everyday interactions between people of differing rank, class, and race are among the construction's lost details. References to such behavior—and evidence of the outrage it could elicit from white men—were documented in official records. White overseers oftentimes defended their brutality by claiming workers ignored them, showed disrespect, or looked at them askance. It is certain that the absolute lack of respect some *tâcherons* showed for their workers stemmed from a sense that the feeling was mutual.

Photographs provide suggestive insights into workers' attitudes as well. Considering the presence of more than one hundred thousand men who worked on the railroad, photos of workers were

A work team, c. 1925.

taken remarkably infrequently. Among the photographs archived by the Batignolles, images of bridges and rails far outnumbered those of the men who built them. In those photos of teams of workers, however, there are often individuals whose expressions suggest the sort of resentment, anger, and disdain that might be expected of forced laborers. Regardless of the uses or intentions of the images themselves, and despite the power wielded by the photographer's gaze, many of the men and boys in them stare squarely, even unforgivingly, at their viewer.

The meanings of facial expressions are certainly not universal. But it is unlikely Europeans could have looked at images of workers, like those in a group photographed in 1925, with their crossed arms and knitted brows, and thought they looked contentedly committed to the project. These men rarely wore the expressions of willing volunteers. European anxieties about authority—a concern expressed repeatedly—belied their own concerns that they lacked it. These anxieties, as well as the feelings of workers, were captured in an insightful report written by Thomann of the Labor Service. The men he encountered on the worksite, he noted, showed "the signs of a profound interior rebellion that we have had great difficulty defeating with persuasion."

If testimony about snide comments and faded photos of disdainful looks provide only a hint at the broad, subtle repertoire workers drew on to assert their individual will, other actions were far more transparent. Men and women often expressed their attitudes in unmistakable terms. White overseers complained that, while unattended, equipment was abused or destroyed, and the carts used to haul dirt were overturned or broken apart. And when vandalism failed to make its point, men refused to work, walked off the job, reported poor treatment en masse to the Labor Service, or simply left and never returned. They were in no way passive victims of overwork and abuse. Thousands of workers challenged authority at their own peril. Disobeying orders risked abuse, punishment, withheld salaries, and even imprisonment. And for those who chose to abandon the construction work entirely, the danger was death. Expressions of resentment or opposition, from sneers to desertion, gave voice to workers' despair and courage, fear and faith.

———

AN INSTRUCTIVE CASE OF workers' disapprobation comes from a much-heralded but ultimately fraught experiment launched by the administration in 1927. To ease the pressure on the African population, the Colonial Ministry in Paris investigated the possibility of building the railway with Chinese "coolies"—indentured workers on a fixed contract. By then, the mortality of African men and women had become well known, and some in France started to wonder if the social disruptions caused by recruitment, as well as workers' apparent lack of resistance to local diseases, would make completing the project impossible. A little over a year later, the colonial inspector Dr. Lasnet began recruiting eight hundred Chinese laborers with the assistance of the governor-general of Indochina. As Chinese indentured workers had helped build other railroads in Africa, under French, German, Belgian, and British direction, the recruitment was widely championed by supporters of the Congo-Océan.

Even in China, however, the reputation of the Congo-Océan preceded the French: as Lasnet recollected, "sinister rumors" meant the French could find no more than two hundred willing volunteers. The Chinese government had officially denied the French request to recruit Chinese workers out of concern for the workers' well-being and because of the absence of a Chinese consular office in Equatorial Africa. Ultimately, Lasnet enlisted the help of a French shipowner in Haiphong, French Indochina, with contacts in Hong Kong who located a large group of about six hundred workers originally destined for Borneo. While it remains unclear how the destination of Borneo, where laborers would have expected to find established Chinese communities, was changed to the Congo, Lasnet got his workforce. In the end, 786 Chinese workers, including four women and a twelve-year-old boy, set off for French Equatorial Africa in June 1929.

The Chinese workers' experience offers a useful point of comparison to that of the Africans recruited for the Mayombe. French officials claimed that they went out of their way to make Chinese workers' transport, housing, and labor conditions as healthy and efficient as possible. The workers were given physicals before departure; most were deemed to be in good health—"muscular and robust"; they were also vaccinated, checked for parasites, and had their heads shaved. They were transported in June to coincide with the dry season in Equatorial Africa to allow them to acclimatize during the time of year deemed "healthier" for workers. Once in Africa, they were divided roughly in half, with 349 stationed in Pointe-Noire and the remaining sent to the Mayombe.

The experience of losing African men and women to malnutrition and disease clearly informed their treatment. Chinese workers were housed in specially built, newly constructed camps with houses, separated from Africans and Europeans by barbed wire. Each worker was assigned a bed, one cover, one mat, one mosquito net, and a shelf for clothes. Food rations, designed to minimize the chance of beriberi, included rice and meat once a day, small amounts of fresh or dried vegetables, bread, tea, and sugar. Lasnet reported that the food was "appetizing and properly served" to the taste of the men. In the inspector's mind, the camps merited emulation elsewhere.

But optimism was short-lived. Aboard ship, a group of recruits beat one of the Chinese overseers so badly, he died of internal bleeding. It was a harbinger of problems to come. Within three weeks of their arrival in the Congo, the Chinese workers were discontented with their situation. In the Mayombe, they soon started refusing to work in what appeared to be an act of organized resistance. On August 6 fifteen refused to work; a week later, forty-seven reported themselves sick in the morning, and eight more refused to work after lunch. The following week two teams refused to clean the camp; workers returned the next day to work but purposefully accomplished little. Within two months the administration was calling Chinese recruitment a mistake. The Chinese were deemed ineffective, too expensive, and a threat to the perceived docility of African workers.

If the administration was caught off guard, it could only have resulted from their woeful ignorance of the resistance mounted by Chinese indentured labor elsewhere. In South Africa in the early 1900s, mine laborers contested poor working conditions by striking, deserting, and rioting against police forces, causing multiple deaths.

In the Mayombe, Chinese workers wasted no time making their dissatisfaction known. Few days went by without some kind of trouble. Scant weeks after their arrival, they complained they'd received no petrol for their lamps, leaving them in darkness after work. Fifty refused to work and encouraged others to do the same. One group of workers threw their tools into a river. When a gendarme threatened to kick them, the Chinese workers gathered to fight. They backed down only when the camp commandant, Captain Houdré, called in a detachment of police.

Lasnet, who undertook an inspection of conditions that led to indiscipline, determined perhaps predictably that the problem lay first and foremost with the Chinese. They were, in his assessment, poor and untrained workers, fundamentally lazy and undisciplined with "bad dispositions"—a striking claim considering how highly touted they had been before their arrival. He admitted that the French were not prepared in a number of important ways. Language was a serious barrier, as only one white officer could communicate with the laborers using a basic "pidgin" Chinese he'd picked up. He also found the men to be "infinitely more difficult to lead than the blacks." With an orientalist prejudice disguised as sensitivity, Lasnet confessed that few officers knew how to deal with "Asians who require such suppleness, diplomacy, and also firmness." Instead, the French were used to the "primitives" who put up with much and demanded little. Africans, he noted, cost little and needed little care; they could be overseen by black *capitas,* while the Chinese required Europeans to direct them.

Despite his lapse into introspection, Lasnet's suggestions for winning over the Chinese workers had little to do with the tact, reserve, and finesse that he claimed overseers in the colony lacked. Instead, the problem was an "insufficiency of repression"; he emphasized the need for intimidation and punishment. The Chinese government had long worked to prohibit the physical harm of indentured workers, so actual brutality was discounted. The alternative was to increase the number of armed gendarmes and overseers, hoping to "dominate" the workers and to "well avoid" future incidents. Of concern was more than just controlling the Chinese: insubordination could spread to the African workers as well.

The administration decided that disobedience was led by a few strong-willed ringleaders who needed to be identified, tried, and punished. The way the Chinese organized themselves "strangely resembled Communist methods," according to one official who was clearly oblivious of politics in China, where Communists had organized labor for years. In response, the French figured that imprisoning men for insubordination and work stoppages might be a "sufficient repressive solution." A number of suspected ringleaders, including one of the translators, a man named Hip, were identified as "dangerous" agitators and removed or imprisoned. Some 172 other workers were deported from the colony in December 1929, less than six months after they arrived. But efforts to punish and deport the alleged ringleaders failed to silence the outspoken defiance of those who remained.

Even with the special precautions taken, the health of the Chinese deteriorated. Two months after their arrival, in mid-September 1929, five Chinese workers had already died—a number that Lasnet deemed acceptable. Conditions worsened in January 1930, when twenty-four men died, and more fell ill in the Mayombe. Morale collapsed; workers in the highlands demanded to be moved to the coast. The Batignolles' Marcel Rouberol concurred; he asked Antonetti to move the workers to the coast, not out of concern for their health but because their continued strikes "threatened progress."

In early February, during Chinese New Year, workers went on strike again. Some 150 claimed illness. The rest went to the construction site, where all but ten of them sat down and refused to work. The work stoppage went on for days, during which the workers complained that they were not receiving their rice ration. Officials again threatened imprisonment, but this time the warning was met with defiance. "I ask only to go to prison," one worker said, "and all my team with me." Later, a disagreement over pay led another worker to throw his money in Captain Houdré's face; the worker danced with joy when he was hauled off to prison. A number of his compatriots asked to be taken to prison as well.

The presence of Chinese indentured laborers offers insight into both the means of resistance available to workers and the French approach to control. Chinese workers contested poor working conditions openly and without reservation. With contracts in hand, they sent delegations to meet with their employers and demand fair pay as well as basic items like towels, cans, sacks, and lamps. They chided the overseers and the translators they mistrusted. They showed themselves to be willing to ask for what they thought fair, including fresh fish and chicken, and even opium. Such outspokenness belied their desperately poor living and working conditions, further exemplified by the preference of many for prison rather than continued employment. But it also suggests what African men and women might well have demanded had intimidation and brutality—prohibited by international agreement in the case of Chinese workers—not been a looming and constant threat.

The French administration ultimately got the upper hand over its Chinese workers; in 1930 it built a prison on the railroad for those accused of insubordinate behavior, forcing Chinese prisoners to work without pay. But it was hard for officials to admit the experiment had failed. Ultimately, they tried to write off Chinese resistance as simply the bad luck of getting a disappointing group of workers. The administration briefly considered contracting another group of Chinese laborers from a different source but scrapped the plan. The large majority of workers returned to China in 1931, at the end of their two-year contract. Many suffered from disease and injuries: from beriberi, malaria, syphilis, mental disorders, amputations, and permanent disabilities. Like many workers on the Congo-Océan, they took a bit of the *chantiers* home with them.

———

DESPITE THEIR SHARED HARDSHIPS, Chinese workers forged no brotherhood with their fellow workers of Equatorial Africa. The two groups looked upon one another with feelings that ranged from incomprehension to mistrust. Little is known of the opinions of African workers beyond the fact that they called the Chinese "rice eaters" and "eaters of pig." On one occasion, Chinese workers threatened to bury some Africans working beneath them in boulders and rubble. Soon afterward two Chinese were apparently murdered, their bodies disposed of in the jungle, perhaps as retribution for bad blood. For the most part, the two groups of workers interacted infrequently. The troubles that officials had with the Chinese in the end increased their esteem for their African laborers. Administrators suddenly appreciated Africans, if for no other reason than that they made relatively little fuss.

For officials to suggest that African workers kept their heads down and followed orders, however, was wishful thinking indeed. African men and women might not have been as overtly or self-confidently demanding as Chinese indentured laborers, but they certainly demonstrated by the thousands their refusal to accept their troubling situation. They had many ways of expressing dissatisfaction and undermining the constant expectation of physically trying work. From early on, company and state officials grumbled that workers feigned illness, wandered off for hours on end, defied assignments or orders, and were slow to return to work after lunch or a midday thunderstorm. Such actions frustrated white overseers and were demonstrations of determination and will in a violent world of forced labor.

Actions have different meanings at different times, but workers saw their own refusal to work as an overtly political act aimed at getting better treatment. Men complained to inspectors about insufficient food rations, poor medical treatment, and excessive labor expectations, especially in task work. And on many occasions, usually after episodes of brutality, entire teams of workers would leave their worksites to file complaints with representatives of the Labor Service. Interviews with former workers in Chad in the 1970s suggested that many men remembered these expressions of dissatisfaction as "mutinies" rather than as mere work stoppages. Local memory celebrated one Sara named Maloum for having rallied his fellow workers against his overseers and successfully winning reforms and getting himself promoted to *capita*. Decades after the completion of the railroad, he was still remembered for his heroism in his hometown.

Most efforts to win reform, however, did not end so productively. Because of the difficulty of winning better treatment, the most fundamental way men and women expressed their intolerance of their overseers was with their feet. A perennial problem for the Batignolles and the administration was "desertion"—a term that reflected the military pretensions of the Congo-Océan as much as it exposed the falseness of workers' "voluntary" presence. In reality, workers who abandoned the construction site likely thought of their decision more as fleeing forced captivity or escaping likely death than as shirking a legitimate duty.

The colonial state encountered the problem of desertion from the moment of recruitment. One of the primary reasons for tying men together by the neck was to keep them from abandoning their convoys. Even when they were not tied, recruits were regularly escorted by armed guards. Whatever the method to keep them together, men still bolted. Desertion among recruits quickly became so common as to be expected. Officials tried to recruit 5 percent above the target quota of workers needed just to cover for expected desertions. But the adjustment failed to make up for the number of recruits who escaped.

The numbers of deserters recounted in various reports speak for themselves. One of the earliest convoys to come from northern Middle Congo evaporated as it headed south; the two *miliciens* who departed with a hundred recruits from Likouala-Mossaka arrived in Mavouadi *alone*. In late 1927 and early 1928, recruits coming from Ubangi-Shari "disappeared" between Bangui and Brazzaville at rates reported to be between 10.8 and 32.7 percent. In February 1928, of the 327 men registered in Bangui, 107 had abandoned their convoys before getting to Brazzaville. In 1929 a contingent of 159 recruits from northern Middle Congo arrived in Brazzaville; of them, 14 died, 16 were deemed unsuitable to work, 70 were sent to the hospital, and 36—some 22 percent—fled. The losses to death, illness, and escape left 23 men, of the original 159, to go on to the construction site.

Once workers were on the railway line, especially for those far from home, the risks of desertion mounted. Leaving the worksite was both the most definitive statement of resistance and the most dangerous decision a man or woman could make. It revealed that workers were no longer willing to tolerate the living and working conditions, and that they had given up hope of expecting reform, justice, or improvement. From the construction site, desertion most often followed instances of physical abuse. As early as 1925, officials linked brutal European oversight with spikes in desertions. The connection with brutality made sense, for workers had to be pushed to the point of accepting the deadly uncertainty of the forest over the inevitability of miserable servitude. It was a choice they made in numbers that confounded and outraged French officials, from the earliest stages of the construction to nearly its completion.

As early as February 1924, Marchand called for a crackdown on deserters, advising the administration to punish not just work-

ers but the villages from which they came. Any worker who failed
to report for work would face swift disciplinary action. If he was
unfound, the worker would be replaced by another from his home
village. A few months later, with desertions—and his frustration—
mounting, Marchand suggested that the family members of absent
workers should be found and held accountable. Doing so would
present "the most effective means of putting an end to desertions."
Marchand's suggestion caught on, with other officials demanding
that deserters be given "exemplary punishments" and that surveillance be increased. But the plan was hopeless: in truth, the administration did not have the manpower to track down hundreds of
fleeing workers through the rain forests of southern Middle Congo.
Instead, desertion was effectively accepted as inevitable.

From an administrative perspective, the exodus of men and
women from the construction site represented a significant loss at
the very moment when disease and death were cutting deep into
the workforce. Officials did not systematically document desertions; rather, references to the departures of men and women, either
alone or in groups, are sprinkled throughout various reports. But
the numbers and consistency were striking. A letter in 1924 noted
that desertions on the worksite at Mindouli had reached a ratio of
one in four. In 1925 an engineer reported that forty-three workers
deserted while being transferred from Kimbedi to Mindouli, about
thirty miles up the worksite. As the circuitous trails along the train
route crossed difficult terrain of dense brush, workers had plenty
of opportunities to disappear into the foliage. The following year,
more than 300 of the 511 workers near Mouyondzi ran away; 38 percent of workers in Kouilou fled; 80 of the 149 recruits in Mandingou
disappeared.

Colonial and Batignolles officials offered many, often contradictory explanations for the thousands of men who abandoned the
Congo-Océan railway line. Predictably few referred to the hardships

of the work or the conditions of life. Equally predictably, they pointed
to a variety of shortcomings particular to Equatorial Africans. In
1924 one high-ranking official dismissed workers' complaints as false
and blamed desertion on the "excessive indolence" that made men
and women avoid work. Five years later, convincing explanations
were still eluding the administration, though some tried homesickness as a promising rationalization. Having rarely strayed from their
native villages, one report argued, workers became "sickly" due to
*nostalgie,* a physical ailment that French doctors, more than a century
earlier, believed to be a real disease. "The workers in any case are
well treated," the report concluded reassuringly. Had there been any
lingering doubts about who was at fault, it stressed that "no complaints have been brought forward by them."

The claim that distance increased workers' homesickness, however, did not coincide with trends. The majority of deserters came
from Middle Congo, no doubt because they were more likely to
make it home, while a very small number of men from Chad abandoned the construction site. In the late 1920s, contingents of workers
recruited in the general region of the railroad lost between 6 and
25 percent of their workforce to desertion. That said, workers from
villages that were by no means nearby also fled. In 1929 an assessment of desertions from Brazzaville noted that the recruits with the
highest rates came from Likouala and Upper Sangha, between three
hundred and six hundred miles to the north. Deserters from far away
would have to rely on a network of people for help. Soldiers, nurses,
former workers, and chiefs of their ethnic or linguistic group sometimes assisted deserters in their escapes. In one case, for example, a
*capita* and a chief pretended to be dead in order to distract guards,
allowing two of their "brothers" to sneak away.

Regardless of the help workers could have received, leaving the
Congo-Océan was an act of desperation and courage. They knew
full well the distances they had traveled on foot and by boat to get to

the railroad. While they had been regularly mistreated and underfed along the way, their voyages to the construction site had nonetheless been organized and overseen by the administration. Workers who deserted in the densely forested, rocky, mountainous Mayombe often did so without knowledge of the people, languages, routes, or geography of the region, or of the river systems, plains, and deserts beyond. Once they were out of the immediate area of the railroad and away from larger towns, their likelihood of encountering Europeans diminished, but their travails would have only begun. Lacking as they did maps, directions, and money for food or transport, their likelihood of escaping and returning home was very low indeed. Settling in a village or in Brazzaville, or escaping into the Belgian Congo, were options; but they posed their own challenges, as men and women would have to find work, housing, and a community willing to accept them. Men and women who fled did so knowing the very real and dangerous risks.

As many deserters were never seen by authorities again, their stories were rarely written down. An idea, though, of the travails of leaving the Congo-Océan was provided by Pierre Contet in 1934. In an article in *Le Populaire,* Contet described an encounter he'd had a few years earlier while visiting Mindouli, the town on the railway known as the center of copper production in the colony. After dinner at a local functionary's home, Contet took a walk with his white hostess. The full moon illuminated the hills and cast rocky outcroppings into dark blue light and deep black shadows. The plateau upon which they walked was deserted and silent: "The spectacle was beautiful." Then suddenly they heard rustling coming from behind a curtain of coffee plants in a nearby field. They approached and found, lit by the moonlight, dozens of bodies lying on the ground, many appearing to be lifeless.

The men had fled from the construction site at M'Vouti, deep in the Mayombe, some 175 miles away from where they now encountered Contet. The men were "thin, emaciated to the bone"; many had "festering wounds"—"large bloody ulcers from which the bland and disgusting odor hit us in the face ten feet before getting to them." Contet was aware that his reader might find his description of the rotting stench to be mere literature. "And yet there is no exaggeration," he insisted, "the living men smelled of decomposing death." Madame C., his hostess, initially recoiled in disgust. The wife of a functionary, she was used to seeing disturbing things; these dying men, however, were a scene of unimaginable horror. She hurried off to find her husband, who called for medical assistance; a doctor who lived nearby came and set up a makeshift clinic in an empty hut.

In the meantime, holding a kerchief to his face, Contet listened to the men's story. "We come from M'Vouti," one explained. "Over there our brothers are dying by the thousands." Seeing so many become weak and sick, they decided to return to their village, near Brazzaville. What began as a group of fifty was now about thirty. And the rest? "Dead, the native responded laconically." They had spent days crossing the rocky plateau and dense forest without provisions. "We want to see our villages again," they told Contet. The next morning ten of the men headed on. Of the twenty-two left behind, three had died during the night, and the rest were "living corpses." The flies swirled around them.

The deserters' experiences in the coffee field where they met Contet revealed the predicament many railroad workers found themselves in. On the one hand, they experienced progressing fatigue, weakness, sickness, and possible death. "M'Vouti, prison among prisons," Contet wrote. "A hell that Dante had not imagined." On the other hand, they feared the unknown countryside, of wandering without direction, money, or supplies, pushed on only by the unlikely chance of getting home. This group of men had chosen the improbability of escape to the certainty of perdition; 80 percent of them were already dead or too sick to continue, still more than

60 miles from home. Other workers who tried to return to homes farther afield—300, 450, or 600 miles from the railroad—held only the slimmest hope of defying starvation, disease, and exposure.

Contet, like other Europeans in the Congo, knew that desertion meant almost certain death. He pointed out that the official mortality statistics that registered a loss of 18 percent of workers—already too optimistic, in his view—did not include the nearly 30 percent of workers whom he believed deserted. Many of them simply disappeared, never to be seen again; the administration never contacted their kin, let alone offered them the back pay earned. For Contet, this was more evidence that the project used the people of Equatorial Africa as though their lives were worthless. The "contempt for human life" could not be clearer, he said, though few were willing to admit it. "Again, this has a name: civilizing!"

Not surprisingly, many officials chose to see desertion in starkly different terms. At fault was not the Congo-Océan or its administration. Poor treatment and conditions were not the causes of desertion. Even many officials who acknowledged that workers suffered from food shortages, poor medical attention, and excessive abuse from their overseers nonetheless blamed desertion on the workers' inherent laziness. For Marchand, insufficient punishment of the "primitives"—who had "never sustained the least constraint on their whims"—left them with "no idea of discipline." It was a common refrain that apparently brought some solace to the white administrators of Equatorial Africa: laziness and a lack of discipline drove men and women to choose a passage across hundreds of miles of unfamiliar and unwelcoming wilderness on the meager chance they'd reach their homes. And it was certainly not the only rationalization to emerge from the Congo-Océan.

# CHAPTER 8

---

# THE MANY WAYS
# OF DEATH

IN EARLY FEBRUARY 1926, the French missionary Père Marichelle was taking a walk in Mavouadi when he came upon a workers' mass grave. Seeing that the tomb had been opened for two new corpses, the priest approached to watch. He noticed something unusual about one of the corpses and asked a nearby *milicien* to remove it from the pit. The body was, much to the priest's surprise, "breathing very well." Marichelle instructed the *milicien* to take the man to the hospital. The priest then went back to Mavouadi, where he'd earlier been invited to join two doctors for a glass of *vin de champagne*. He recounted his story, which confused the doctors; they insisted that procedures would make it impossible for such a thing to happen. An official investigation was attempted a few weeks after the fact, but like so many others, it failed to either corroborate or disprove the priest's story.

Factual or not, Marichelle's story captured a certain truth about mortality on the Congo-Océan. Sickness and death were ever-present. Malnutrition, disease, exhaustion, and injury made for a mortal blend that took lives indiscriminately and relentlessly. The dying far outnumbered the medical personnel available to care for them, and far outstripped the administration's effort to keep track of them. Days sometimes passed before the dead were even noticed. It's not hard to believe that a living man might have been buried alive.

Although Antonetti and other defenders of the railroad insisted there was no link, workers' conditions on the Congo-Océan—including their recruitment from hundreds of kilometers away—certainly contributed to the exceptionally high morbidity and mortality rates. The large majority of men and women who died once they were in the grip of *la machine* were ultimately killed by disease. But the routine of nine-to-ten-hour workdays and six-day workweeks, the punishing demands of porterage, and the psychological fatigue of forced labor did nothing to strengthen immune systems. Health conditions on the Batignolles' western end of the line were worse than on the stretch of railway built directly by the state. But as the inspector Dr. Pégourier noted in 1926, even the latter were only treated with "relative humanity, to say the least."

Doctors on the construction site approached the health of the workforce in terms that included all aspects of their bodily experiences: as one physician put it in 1925, it was essential to take "all measures likely to minimize fatigue and to augment physical resistance." Measures included protecting people from germs by promoting washing hands, burying feces, and burning garbage; but they also aimed to minimize porterage, improve food and living conditions, and establish progressive training and work schedules to avoid exhaustion. It was a pressing issue: doctors were only too aware that a significant percentage of the workforce would not live to fulfill their obligation. In 1925 one physician estimated that 22.5 percent of workers would die annually. The estimate was low for new recruits from the north: between 1925 and 1928, many of those contingents lost more than half their number.

If the colonial regime was unprepared to recruit and transport thousands of workers, as well as house and oversee them, it also had great difficulties feeding and protecting them from disease. Just as the decisions that pushed the project forward made brutality at the point of recruitment and overcrowding on riverboats more likely, they also made malnutrition and untreated illness realities for many thousands of workers. Multiple inspections and investigations studied and identified the causes of endemic illnesses and epidemics. But scientific knowledge was very slow to improve the situation. Doctors struggled to treat, cure, and, in some cases, even diagnose a range of maladies. Meanwhile thousands of men and women from the north continued to pour into worksites. For their part, administrators, Antonetti foremost among them, refused to accept that recruitment from the north had been done hastily and without proper planning.

The recommendations of doctors and inspectors were often adopted, in theory, as policy. The writing of regulations, however, did not mean that practices changed on the ground. Significant improvements in the health of workers took years to achieve. Alarmingly high rates of morbidity and mortality improved only toward the end of the 1920s, seven to eight years into the construction. Real relief coincided only with the end of major construction in the Mayombe. All the while, the work moved forward without any suggestion from the Batignolles or the colonial administration of slowing or postponing construction to improve conditions and, thus, spare lives.

The deaths of men and women on the Congo-Océan made plain the extraordinary weakness of the colonial regime. The bodies of starved, sick, and dying workers were manifestations of the ineffectiveness of French rule. Despite the colonial government's claims that it was taking the necessary steps to foster the lives of its workers, increasingly it was forced to face the reality that it could not keep them from dying. Worse still, the very effort to modernize this corner of the empire exposed thousands of men and women to bacteria and viruses—from shigella bacillus bacterium to human immunodeficiency virus (HIV)—that had thrived for perhaps millennia in places few people ever went. Defenders of the railroad often championed its capacity to open central Africa to commerce, industry, and civilization. But the Congo-Océan also opened the region to diseases that proved to be more devastating than its defenders could have ever imagined.

———

IN ADDITION TO TOLERATING abuse, strenuous travel, and inclement weather, workers' main challenge was simple survival. Shortly after recruits left their villages for the trek to the railroad, hunger and thirst set in. Major posts had supplies to feed men, but recruits had to cover dozens or hundreds of miles between posts that sometimes did not. In August 1925 a report from a Mr. DuBois laid bare the colony's lack of preparation for expanded recruitment in the north. Reflecting on the "lamentable conditions of this too hasty recruitment," DuBois decried the fact that no one had studied the means of resupplying the recruits. "It was essential to create *beforehand* some 'granaries' of food at different stopping points, along the men's route, and on the different worksites," he wrote. "On this issue, nothing has been done, organized." Instead, the colonial state had recruited thousands of men and put them on the trail to the railroad "without worrying about feeding them."

Recruits were meant to be provided with seventy-five centimes per day to purchase food along their route and even after arriving at the construction site. The problems with that plan were legion. As observers noted, recruits had little idea of the value of the money given them, making them easy prey to cheats. Albert Londres found that merchants gladly exchanged a handful of iron for the workers' cash. Workers, he pointed out with his deadpan irony, don't eat iron. A more profound problem was that there was little food, if any, for them to buy. With the exception of larger trading posts, the countryside lacked markets, as few villages produced excess food for sale. Without canteens, potable water was equally scarce, especially during the dry season. The result was that many recruits did not eat or hydrate sufficiently. Some, according to the colony's own reports, did not eat at all.

Even if the system of payment had worked, seventy-five centimes was a meager amount of cash to feed oneself sufficiently, as the administration knew. In 1923, to offer a point of comparison, the Batignolles estimated that their white agents needed 150 francs per month to secure food, that is, more than six times the amount provided to recruits. Between the insufficient opportunities to procure food and paltry allowance to buy it, the consequences for recruits could be devastating. In 1925 nearly 20 percent of recruits from northern Middle Congo were deemed "emaciated, skeletal" upon arrival. As a result, health inspectors insisted repeatedly that recruits be provided with food, not money.

Defensive of their recruiting procedures, some officials pushed back, insisting that recruits were given adequate food, which they had mismanaged. In 1925 the acting lieutenant governor of Middle Congo, for example, reported that recruits had been supplied with all they needed for their travels before departing to Brazzaville. That system, however, left recruits having to carry heavy supplies, sometimes for hundreds of miles. They chose instead to eat all they could as soon as they received it, not knowing, of course, how long their transit would take. Others sold or gave away their supplies. But debates between officials only highlighted that the procedures for efficiently distributing food had clearly not been determined, nor would they be for years to come. Dispersing cash rather than food was ultimately deemed best, and it remained the norm until the late 1920s.

Once they were on the construction site, workers were often expected to eat foods they had never tasted. As with many aspects of equatorial societies, the culinary traditions of recruits varied greatly.

Manioc had become a common staple across most of the colony. It was easily grown, even in poor soil, and could be harvested as needed twelve to thirty months after planting. Two centuries of slave raids had made it an important crop for populations that had to resettle at a moment's notice; once the threat passed, they could return later to find their manioc waiting for them. Beyond manioc, however, staples varied. Recruits from the forests near the railroad typically consumed plantains, small game, and manioc greens. Workers from the savannahs of Ubangi-Shari and southern Chad were used to diets that included millet, maize, vegetables, and chicken. Some groups, like the Banda, supplemented their diets with hunted meat; riverine groups fished. Workers from the edges of the Sahel ate more meat and dairy from herd animals such as cows, goats, and sheep, as well as a range of fruit and vegetables.

Regional eating habits were ignored on the construction site, replaced by a single diet that was as limited as it was apparently unappetizing to most workers. In the first years of the construction, when the bulk of workers came from the region relatively close to the railroad, men often supplemented their rations with familiar foods they found in the bush. No large animals lived in the Mayombe; but small animals and insects could provide supplemental protein. But even those from the equatorial forests were often hungry. Visiting the railway line in 1924, one doctor saw the "famished wretched" workers who lived in "a state of deplorable physical decline . . . bordering on starvation." The railway project created a state-sanctioned famine.

The situation did not improve in 1925, when some two thousand recruits arrived from the far north. Unable to provide the quantities of manioc needed, the colony relied instead on rice as the centerpiece of workers' diets. Rice was not a typical part of most equatorial diets, and officials acknowledged that workers did not like it. Plentifully produced in French Indochina, relatively easy to import and transport, and slow to rot, rice was forced on workers, despite the fact that workers greatly preferred manioc. A rice ration was set initially at 500 grams per day, though it was raised to 800 grams months later. Rounding out workers' daily diet were, according to regulation, 20 to 30 grams of palm oil, and 100 grams of dried fish or 200 grams of fresh meat. Plantains, peanuts, and other foods were sometimes distributed; rare were fresh vegetables and fruit. Workers were encouraged to supplement their rations with foraged or purchased goods, though opportunities to acquire more food were limited.

Rations were more optimistic goals than reality, as workers regularly did not receive part or all of the food promised. The most fundamental problem was that there was simply not enough food for the thousands of workers. Food production along the railroad had never been robust due to the traditionally sparse population. In the first years of construction, villages tried to keep up with the demands of the railroad, providing the colony with thousands of pounds of manioc, as well as thousands of men and women for porterage and work. It was, said one official in 1923, "a heavy task," one that ultimately was too much for villages to handle, leading to an exodus of the population, especially to the Belgian side of the border. By 1925 the colony had begun making a concerted effort to grow its own crops, developing farms in the region. The soil was arid and sandy, and even the sweet potatoes and banana trees died. "As for crops," Thomann lamented with a certain poetry, "we have, for the moment, reaped only woes."

With dwindling local supplies, the project relied overwhelmingly on products imported from outside the region, requiring a huge amount of food to be procured and moved. Shipments and deliveries were sporadic and often inadequate. Shortages of rice forced the ration, raised in 1925 to 800 grams, to be lowered again; palm oil, a common ingredient in many central African dishes and an important source of dietary fats, was not regularly distributed until well into 1926. The first fresh meat arrived only in December

1925, and the quantity revealed the severe shortages: two cows were slaughtered for distribution to, in theory, more than five thousand workers. The following month was somewhat better; fourteen cows were slaughtered, but that quantity still failed to provide substantial sustenance for the workforce.

The quality of the beef also left much to be desired. One health inspector saw twenty-four cows arrive in Pointe-Noire from Cameroon that were of good size but very thin. By the time they made it to the Mayombe, they had lost all their fat, and without pastures or feed, there was no way to rehabilitate them. The farther they went, the more the cattle suffered: those that arrived in M'Vouti could be saved only by butchering them, their meat already without fat and of poor quality. That said, beef was welcomed by Europeans and workers alike as an uncommon delicacy.

The arrival of beef—two cows one month, more another—was typical of the piecemeal nature of food supplies in the Mayombe until the late 1920s. One inspector, Dr. Boyé, lamented the myriad challenges in feeding the men: no local cultivation due to depopulation, delayed imports, poor communication, and a climate that could make even rice rot—"all this represents the elements of a singularly arduous problem." Efforts to feed people and combat malnutrition, then, were slapdash by nature. In 1925 one medical inspector called for the distribution of peanuts to all workers. Peanuts were rationed from August until December 1925, when they ran out. The peanut ration was then replaced with shea butter (*beurre de karité*), made from the oil of the nut of the shea tree. Its usefulness was mixed, however, as only workers from Chad were willing to eat it. The following February six tons of peanuts arrived from Senegal, allowing for a ration of seventy-five grams a day, a good supplement to the rice ration, but one that would have lasted less than a month if distributed regularly to thousands of workers.

Shortages were compounded by the problem of distribution.

Most food, especially in the Mayombe, was packed in on the heads or backs of porters, who themselves represented more hungry mouths. The quantity of supplies needed to feed—or underfeed—thousands of people was remarkable. Some 15,000 kilos (about 33,000 pounds) were needed every four days to provide the manioc ration. Carried along the circuitous goat paths of the forest, many supplies were damaged or rotted before being distributed. Sacks of rice were easily punctured; grains slowly leaked out as porters made their way to delivery. The loss of foodstuffs in transport was passed on in the distribution.

On the worksite, rations were meant to be handed out in two weekly distributions. Semistarved, many workers ate most or all of their ration upon receipt, leaving them with nothing to eat for the following days. Compounding problems further, certain rations, like manioc and millet, had to be processed into flour before they could be consumed, a task traditionally done by women. While women on the site regularly cooked for men, the significant gender imbalance made it impossible for women to cook for every man, leaving many men to fend for themselves. As many men did not know how to make manioc flour, they consumed it in ways that made it difficult to digest.

Food shortages caused discipline problems, fueling corruption and turning men against one another. Porters took portions of food from what they carried. Once at the distribution points, workers' rations were also shared among what one doctor called "all the parasites, friends, women, etc." who sometimes passed through camps. Intimidation and connections had their effects, with "the strongest taking the lion's share." Hunger drove workers to steal from one another and to raid villages they passed for food. Women sold their bodies for food; men sold their wives for food. Empty stomachs caused despondency and, for the administration, worse: desertion.

Officials from the Batignolles, men who rarely complained about the treatment of workers, found the shortages of food and the failed deliveries to be worthy of protest to the administration. In November 1925, Rouberol reported to the governor-general that multiple shipments of supplies had delivered only rice and salt for the men, notably excluding smoked fish and palm oil. Lest one think Rouberol was driven by empathy, he was quick to state that the company's concern was in productivity: hunger led men to "threaten not only to quit our worksites" but to dissuade potential "volunteers" from joining the workforce as well. The complaint had a larger point to make: should hunger slow the railroad's progress, the Batignolles would not accept blame.

Despite recommendations to prepare food daily for workers in centralized kitchens, the administration repeatedly refused. The reasons were mainly logistical and financial. But experience also showed that Africans often refused to eat food served to them, voicing deep mistrust of Europeans. Workers' wariness also included suspicion of the food rations and medical care provided on the railroad. Their reasons for not eating clearly varied. Some apparently could not bring themselves to eat food that seemed entirely foreign to them. One official was shocked to see "emaciated" men from northern Middle Congo refusing to eat rice and dried fish served to them. "Even our starved skeletal invalids refuse it, all demanding manioc and meat."

Testimonies suggest that it wasn't mere preference that kept workers from eating: the food was often inedible. According to Marcel Homet, the food distributed was disgusting even as late as the 1930s. An old European showed him the dried fish—called *pongo*—that workers were meant to eat. "What rot!" Homet said that it stank from a fifty-foot distance. He was perceptive about the relationship between the rations the government promised to workers and the reality on the ground. The official journal of the colony, he noted, gave luxurious descriptions of the food and calories meant for workers. "And in accordance with these prescriptions, rotten manioc, rancid palm oil, and spoiled fish (when by chance it arrives) were served to the workers." And workers died by the thousands. Homet asked a railroad administrator why they didn't resupply the "poor devils" more effectively. The answer was simple: "Because it cost less to find new workers."

————

ALL THE INCONSISTENCIES IN delivering, rationing, and preserving food make it impossible to estimate how much people on the railroad actually ate on a regular basis. Indeed, the one certainty was that little was "regular" about the Congo-Océan diet, especially before the late 1920s. Geography played a role; workers closer to Pointe-Noire or Brazzaville benefited from being near food storage areas and likely ate more regularly. But people worked at different tasks; bodies reacted differently to scarcity, no matter where they were posted. The most poignant way to understand the misery of hunger is through individual accounts. But European descriptions of food shortages and disease rarely recorded the voices of afflicted workers. Other ways of measuring hunger and thirst, then, must be considered. "The experience of suffering," as the medical anthropologist Paul Farmer has rightly warned, "is not effectively conveyed by statistics and graphs." And yet that was how medical inspectors on the Congo-Océan tried to understand hunger.

The ration, an attempt at that statistical measurement of nutrition, is a place to start. Meant to be a guarantee of a base level of consumption, the ration in the mid-1920s was government-subsidized deprivation. The 500 grams of rice, 100 grams of dried fish, and 30 grams of palm oil promised in the ration provided workers about 1,300 calories per day. The additional of 300 grams of rice—to meet the 800-gram ration—would have added another 400 calories. In addition to being paltry sustenance, the diet was known to be "very poor" in terms of vitamins. Comparing cases of deprivation is tricky, especially with diets where levels of activity and access to supplemental food can differ greatly. But to put it into some perspective, the Congo-Océan's rations were very meager, comparable historically to the sustenance provided in concentration camps and regions torn asunder by war.

By 1930, some nine years into the construction, food distribution had become more regular, especially with the completion of the service road into the Mayombe that made shipment of foodstuffs considerably easier. By that point, Antonetti had also started encouraging more women to come to the worksite with their husbands. More knowledgeable about food preparation, women often cooked for teams of men, a more efficient division of labor than leaving food preparation to individual men. Women's meals were likely tastier and were somewhat healthier, as women knew how to prepare certain foods, like manioc and millet, in ways that maximized their nutritional value. Deliveries became more regular and their preparation improved, but rations were not substantially expanded.

Calories have a political history all their own and can say as much about a society's ideal of consumption and productivity as about personal health. But in the case of the Congo-Océan workers, French officials clearly knew that the rations provided were well below their physiological needs. If the administration's own doctors' concerns weren't worth listening to, perhaps the international community was. In 1925 the International Labor Organization in Geneva launched a general study of worker nutrition. The ILO set the standard caloric intake for working adults at 2,500 calories. On its low end, then, the ration on the Congo-Océan was just over half of the international standard. Doctors on the railroad line knew the ration was dangerously small and told their superiors. One doctor touring in 1924 found that everywhere he asked men about their food, "the response was invariably the same: we are insufficiently fed."

The same did not hold for the Europeans on site. The white men working on the Congo-Océan regularly complained of the physical hardships, but their experience of hunger could be remarkably different from that of workers. The European *surveillants* and *tâcherons* who lived next to their men on the site grumbled about the lack of the staples they were accustomed to, such as bread or flour. Malnutrition per se was never considered an issue, but the unappetizing and inconsistent diet of European workers was deemed a cause of their general malaise and poor health.

The professionals of the Batignolles in Pointe-Noire had even less to complain about. Regular shipments from Europe tried to emulate, if not entirely equal, the flavors of French cuisine. Orders of supplies from the Batignolles in Pointe-Noire read like Parisian shopping lists, albeit with bulk quantities: two thousand liters of red wine, one thousand liters of white, and two cases of sparkling wine; a case each of veal, beef, and pork conserved in fat; sausages, salmon, and sardines; kilos of green beans, Brussels sprouts, and carrots; Gruyère, two cases of jam, coffee, and sugar. Neither rice nor dried and salted fish was procured for the men of the Batignolles.

For workers, food did not nourish a nostalgia for home; rather, malnutrition had predictable metabolic consequences. What was the actual physiological impact of semistarvation on workers? Malnourished bodies break down muscle protein to produce glucose. Edema, or the swelling of joints, is a common side effect of semistarvation and indeed was reported in workers. The stress on the body caused by a lack of sustenance actually speeds up the metabolism that can precipitate cachexia. For men and women who do not have excess body fat, severe undernutrition can be excruciating. As one researcher has written, "non-obese individuals often fixate on food and suffer from severe, prolonged, and distracting hunger pain." If the short-term effects were bad, prolonged hunger in young people—of which there were many on the Congo-Océan—can cause long-term developmental problems, stunting growth and brain development.

Beyond the breakdown of bodies, surprisingly few scientific studies exist of the wider effects of prolonged semistarvation and undernutrition. One influential scientific study, however, offers some sense
of what workers on the Congo-Océan endured. In 1944 Ancel Keys,
an American physiologist, led a team that studied the impact of semistarvation of the human body and mind. Known as the "Minnesota
experiment," the study limited thirty-two young men (the average
age was twenty-five), all conscientious objectors of the Second World
War, to a diet of about 1,570 calories per day for six months. The aim
was to better understand not only the impact of semistarvation on the
body but also the best ways to treat undernutrition and rehabilitate
starved populations, a particularly important issue considering the
millions who suffered food shortages in war-torn Europe.

Many aspects of the Minnesota experiment differed greatly from
the conditions in Equatorial Africa, most notably, the nutritional
histories and the diets of those affected by semistarvation. But the
findings of Keys and his colleagues remain instructive. In particular,
the experiment documented the profound mental and psychological
consequences of semistarvation. Severe hunger renders people irri-
table, moody, and apathetic; their sensitivity to some stimuli, especially noise, increases, while mental acuity, self-discipline, and drive
decline. The mental health of workers, too, was jeopardized by privation; prolonged hunger caused or worsened depression in men and
women far from home. Laziness, an inability to focus on tasks, and
a lack of discipline were standard complaints that white overseers
made about their workforce. White men in the Congo, however,
blamed not starvation but racial inferiority.

Photographs taken of sick recruits showed Batignolles executives
back in Paris the impact that malnutrition and disease could have,
exposing men's ribs and rendering limbs little more than skin and
bone. Again, such images were not taken to elicit indignation—
or if they were, they did not succeed. Rather, for European engi-

A team of sick workers at Kilometer 61, Mavouadi, June 1925.

neers far away, emaciated bodies were simply part of the equation in Equatorial Africa. They undoubtedly concurred with doctors who explained workers' physical deterioration as a sign of their failure "to adapt" to the new environment they found themselves in. There was no suggestion, however, that recruitment from the north should stop. In fact, the Batignolles regularly complained that they had insufficient numbers, pushing the colonial state to recruit more workers and from farther afield.

The pain of extreme hunger is notoriously difficult to describe, and few witnesses to the Congo-Océan tried. Even in the hands of the most talented writers, hunger often blends with metaphor. Many men and women on the railway lived with hunger, not simply constantly at their elbow as they worked but also when they woke in the night. In the darkness, as Richard Wright wrote of his own family in the 1920s, hunger stared at them gauntly. But reports and accounts from the railroad often tried to account for semistarvation in plain terms, which often provided eloquent testimony. In 1926, for example, a concerned doctor described an adult male worker whose "remarkable state of emaciation" had reduced him to 35 kilos (about 77 pounds), less than two-thirds his original body weight. The man had lost nearly 20 kilos (44 pounds) in a matter of weeks. All that remains to history of this man who watched helplessly as his flesh wasted away is the administrative identity given to him by the French: No. 8846.

Albert Londres witnessed firsthand the breakdown of food distribution in a contingent of men making the two-week trek from Brazzaville to the Mayombe. Only half of the time, he reported, did supplies arrive on schedule to feed the workers. For the other half, "the convoys wait in vain for millet and salted fish." Sometimes the convoy encountered large stocks of supplies, but the *gardien* didn't have the authority to let them eat, since regulations had not "provided for workers to be hungry at this stage. *Hunger! Hunger!*—this tragic word rose all along the route." The men moved along the line, disintegrating until the orderly convoy looked like "a long injured snake" of dragging men, their *capitas* driving them on with the *chicotte*.

Londres was not the only witness to blend discussion of hunger with images of physical violence. Descriptions of starvation on the Congo-Océan often alluded to other, more visceral forms of torment. One worker remembered his experiences of malnutrition as indelibly linked to memories of brutality: "Tortured by hunger, pick or basket in hand, under the eye of a sadistic and blood-thirsty foreman, many workers got dizzy, fainted, and hurt themselves." Deprivation, fatigue, and abuse were the mantra of the railroad worker.

———

IN ADDITION TO SUBJECTING workers to the pain and fatigue of hunger and thirst, inconsistent diets caused or exacerbated other health problems. Within weeks of their arrival at the Congo-Océan, recruits from Ubangi-Shari started to develop beriberi, which quickly became the second leading cause of death on the worksite. Unknown in Equatorial Africa before the construction of the railroad, beriberi is caused by a vitamin B1, or thiamine, deficiency; early studies found it to be common in people, such as Japanese sailors, whose diets relied on large quantities of polished, rather than unprocessed, rice. As workers started arriving from the north, polished rice, which lacked the more nutritious husk of the grain, became the bulk of the daily food ration, replacing manioc.

In 1926, when Dr. Boyé wrote an extensive report on health conditions, the exact causes of beriberi were still, as he put it, "undetermined," but it was known to be linked to diet and vitamin consumption. Boyé witnessed beriberi arrive with the first recruits from Upper Congo, the Sangha, Ubangi, and Chad. In May 1925 he encountered only two cases of beriberi; a year later, after recruitment from the north increased, there were 97. The number of monthly deaths grew from 0 to 25, taking 116 men's lives over the course of the year. Georges Lefrou, arguably the most committed doctor on the Congo-Océan, working largely alone across the entire construction site in the mid-1920s, reported that men suffering from the disease experienced weakness and generalized edema as well as blurred or impaired vision. Less common on the railroad were the neurological disorders often associated with the disease. He correctly encouraged the distribution of peanuts to combat beriberi, but shipments were irregular.

Malnutrition was also related to a condition simply referred to as *misère physiologique,* severe physiological distress. It was a kind of umbrella diagnosis of men rendered weak and emaciated, unable to work or move, their muscles deteriorating. Such cachexia was associated with a host of illnesses, from beriberi to ancylostomiasis (hookworm) and other parasitic diseases. If health inspectors were uncertain about the various causes of *misère physiologique,* they generally agreed that it was connected to poor nutrition.

There is no doubt that malnutrition cost lives. Official morbidity and mortality rates must be taken as rough estimates at best. Autopsies were not possible on the site, so causes of deaths, when recorded, were based on symptoms and educated guesswork. Nonetheless, doctors' estimates offer some idea of the prevalence and impact of diet-related illnesses. In the mid-1920s, Dr. Lefrou estimated that beriberi and malnutrition accounted for about 13 percent of deaths. As about a quarter of deaths between 1924 and 1926 were of causes unknown or undiagnosed by health workers, the number who perished directly from a lack of food was very likely higher.

Colonial inspectors insisted that these deaths were preventable, but that men and women could not be saved by medicine alone. Rather, the cure for beriberi, malnutrition, and semistarvation could only be delivered by administrative changes. Workers, in other words, died because of the colony's inability to organize and distribute the supplies needed to keep them adequately fed. As Dr. Boyé noted in 1926, prophylaxis was impossible from a medical point of view; the problem was of a "strictly administrative nature"—a point he chose to underline. Beriberi and malnutrition, then, were structural problems more than medical ones. Beriberi, Boyé continued, "is a disease of food deficiency; the doctor can do nothing to avoid it. The administrative authority can do *almost everything.*"

OTHER DISEASES HAD MORE complicated relationships to the building of the Congo-Océan. Most recruits who died of disease on the railroad did so only because they had been brought to the region in the first place. Men and women from the north were exposed to bacteria, viruses, and other dangers endemic to southern Middle Congo, against which they had limited natural immunity. Cramped, close living quarters were ideal conditions for the spread of maladies, from airborne illnesses to venereal diseases. And again, hunger played a role. The effects of the most common bacterial infections that workers faced, especially respiratory and intestinal ailments, are worsened by a lack of calories. Hunger renders people more susceptible to accidents and fatigue, both of which, in a climate like the Mayombe, could be life-threatening.

Dangers on the construction site ranged from the widespread, such as skin ulcers caused by going barefoot in the inclement forest, to the infrequent but potentially deadly, like bites from poisonous snakes. But the most serious ailments read like an encyclopedia entry of tropical diseases. Hookworm, microfilariasis, and intestinal parasitic ailments as well as malaria and undiagnosed fevers were common. A yellow fever epidemic in Matadi was contained in early 1928 after energetic efforts by the Health Service. Respiratory ailments, including lung and bronchial infections and pneumonia, were particularly virulent in the mid-1920s. Lefrou estimated that between 1924 and early 1926, 50 to 68 percent of those who had respiratory illnesses died from them. In 1927 there were nearly 500 cases of respiratory infections in Pointe-Noire out of a workforce of 1,700; 195 men died from them. African trypanosomiasis, known as sleeping sickness, a common disease in parts of Equatorial Africa, causes severe fatigue and damage to the central nervous system; thankfully, it was fairly rare on the Congo-Océan. Cases were limited to recruits who'd been exposed before arriving on the line.

The "most murderous" killer of workers was bacillary dysentery, a particularly virulent and highly contagious form of dysentery caused by the shigella bacillus bacterium. Recruits' country of origin determined the potential danger of the disease. Endemic in the region around the lower Congo River, bacillary dysentery took relatively few lives of workers from southern Middle Congo, where exposure earlier in life seemed to provide some immunity. For the thousands of recruits from the north, however, the story was very different. Transported to the railroad down the Congo River, those from northern Middle Congo, Ubangi-Shari, and Chad sailed into the infected lands.

In the mid-1920s, about a third of all deaths on the railroad were attributed to the disease. The morbidity and mortality rates reflected the geographical differences of the workforce. Less than 1 percent of men from around the railroad site died from dysentery, while nearly 10 percent of some contingents of workers from upriver and the north perished. Based on a nine-to-twelve-month study of recruits from the north, Lefrou found that as many as 15 percent suffered from the disease, and between 5.1 and 8.5 percent died from it. One good trait of the disease is that migrants from the north who survived the first six months were less likely to be infected later; dysentery moved quickly.

Dysentery became a reality of everyday life on the *chantiers,* as men and women experienced it both firsthand and as witnesses to the sick around them. For many people living in the developed world today, dysentery can seem like a distant ailment relegated to the history books. But for the afflicted, as a late nineteenth-century health guide for travelers to central Africa described it, the disease was violent and devastating. Those afflicted were struck by intense abdominal pain, vomiting, and pungent diarrhea, which was often bloody from the beginning of the illness. The blood appeared in either "thick streaks or more or less closely mixed in the liquid." In severe cases, patients discharged "strips of membranes" from their intestines.

As the disease ran its course, the sick in the Mayombe could have burning fever and up to eighty bouts of diarrhea a day. The only defense against the illness was calomel, a toxic purgative derived from mercury chloride and a dangerous poison in its own right. Many suffered from tenesmus, a cramping pain in the rectum, and bent double as they struggled to evacuate their bowels. The terminally ill lay prostrate, collapsed, and unable to move. In the final days of the disease, the afflicted also discharged pus; in such cases, the diarrhea lost its fecal smell and took on an odor described as "bland and nauseous." For the dying, the bowels calmed, but dehydration set in, leading to cramps, a burning throat, delirium, and organ failure.

If the worksites lacked amenities for the living, they offered even fewer for the dying. In the forests of the Mayombe, without running water, decent housing, and sufficient medical facilities, the misery was only compounded. During particularly bad outbreaks, medical facilities were overrun. One can only imagine the smell and sight of camps and understaffed medical stations overrun with dozens of dysentery patients. In 1925, during one such episode, Lefrou reported that many refused to go to the bush to defecate, instead "preferring to defecate around the hut"; the "sandy soil" of the huts became a veritable petri dish of bacteria. If the epidemic continued, one official encouraged the construction of isolation wards with cement floors to ease the washing away of filth. But these facilities were very slow to be built.

Many workers refused to take medicine; they had what one administrator called, "a superstitious terror of the Whites' hospital"—a colonial anxiety not at all limited to Equatorial Africa. Some remained in their huts, where they became reminders to other workers of what their futures might hold. Others "hid themselves," seeking refuge "in a corner of the bush"—a euphemism for men and women who'd lost such hope that they chose to die alone in the dark mud of their own filth, blood, and bile. The administration considered workers who did this to be "deserters," and as one administrator admitted, they "escaped our statistics." Dying alone and lost from home, these workers faced the final indignity of literally being ignored by their colonial government, expunged from records, overlooked by administrators and the archives they compiled.

WORKERS ON THE CONGO-OCÉAN also died from a disease that had never before been diagnosed by doctors. Medical researchers have since speculated that the ailment might have been caused by HIV-1, the virus that causes AIDS, a disease that would be identified and diagnosed only when it became a global pandemic in the 1980s. For years, teams of zoologists, doctors, evolutionary biologists, and journalists have studied and debated the origin and histories of the various strains of the human immunodeficiency virus (HIV). The story, with its globe-trotting scientists and unlikely discoveries of decades-old blood samples on dusty clinic shelves, can read like a murder mystery. Disagreements about the origin of HIV-1 and the spread of its most deadly group M ("Main") have at times been volatile. But a number of compelling scientific studies about the origins of the disease make it essential to consider the possibility that some men and women on the Congo-Océan railroad were infected with and died from an early form of AIDS.

Current scientific research holds that HIV-1, the type of the virus that has killed millions of people worldwide, originated as a simian immunodeficiency virus (called SIVcpz) found in the chimpanzee subspecies *Pan troglodytes troglodytes*. It is unknown how long SIVcpz has existed in *Pan t. t.*—perhaps centuries—but at some point it was passed from chimpanzees to humans. This transmission likely occurred on a number of occasions, but one occasion, the spillover from a single chimp to a single human, resulted in the HIV-1 group M that would eventually become a global pandemic. Cross-species transmission had to have resulted from an exchange of blood or saliva. In central Africa, as people hunted and ate primates, including chimpanzees, a hunter was likely the first to come in contact with a chimp contaminated with SIVcpz. Exposure to blood could have plausibly occurred during the carrying or butchering of a chimp's carcass. A lesion on the shoulder of a hunter hoisting the animal's body, or a cut on the hand of the person who worked with raw, bloody meat could have caused a transmission with catastrophic consequences to take place.

Studies using viral sequencing and the methods of evolutionary biology have determined that this spillover occurred, at the latest, in the early twentieth century, in the region where *Pan t. t.* live, most likely in southeastern Cameroon, near the border with Middle Congo. From there, the virus moved south in human carriers, arriving in the region of Stanley Pool probably in the 1920s. It was clearly present in Léopoldville and Brazzaville in the 1930s. In other words, the movement of the virus coincided temporally and geographically with the movement of tens of thousands of recruits from northern Middle Congo and Ubangi-Shari to the Congo-Océan.

The question, then, is whether men and women on the railroad showed possible signs of AIDS. One microbiologist, Jacques Pepin, has found that in the early 1930s a connection between HIV and deaths on the railroad was potentially, if unknowingly, documented. Pepin followed the career of a remarkable doctor, Léon Pales, who arrived in Brazzaville in 1931 as the colonial surgeon. Shortly after his arrival, Dr. Pales found himself with time on his hands in a relatively slow outpost. Long fascinated with pathology, he started conducting autopsies on Congo-Océan workers, using the Pasteur Institute's laboratory, to better understand the bacteria behind common respiratory and intestinal ailments. In a study primarily of tuberculosis, Pales conducted autopsies on 124 workers from the Congo-Océan who had come from across the colony. He found that thirty-five of the workers died from what he called "cachexia of unknown origin." He set out to pathologize this disease in an effort to better understand the myriad causes of death in Equatorial Africa.

In twenty-six of the autopsies he completed, Pales identified a malady that neither he nor other medical officials had ever previously diagnosed. The men showed a number of similar symptoms. First, they all suffered an extreme form of wasting that left their bodies weighing less than 35 kilos (around 77 pounds) and inevitably resulted in death. While severe weight loss was not uncommon on the Congo-Océan in the late 1920s, by the time Pales conducted his autopsies, food distribution had improved dramatically and diets had stabilized. The condition was so striking that Pales dubbed it "Mayombe cachexia," as all the workers he autopsied had worked in the region.

The bodies that Pales examined had also suffered from chronic nonbloody diarrhea (testing negative for common forms of dysentery) and yet retained healthy appetites. Hospitalization in Brazzaville had not improved their condition, even with attentive care and improved diets. Many of the deceased also suffered from cerebral atrophy and enlarged mesenteric lymph nodes (lymphadenopathy), with no sign of tuberculosis bacillus. Autopsies and bacteriological tests were unable to determine the causes of the wasting. One thing was certain: the malady was undoubtedly contagious, considering the concentration of cases.

Like AIDS, the "Mayombe cachexia" suffered by Pales's autopsied workers caused physical and psychological misery. Hungry but unable to gain weight, the men ate regularly, only to see the nutrients pass through them quickly. In time, they were wracked by the symptoms of starvation victims, their bodies unable to process protein. The metabolic response was to break down tissue, including organ and bone, causing extreme wasting and physical weakness. Many of the men Pales autopsied had suffered from cerebral atrophy, which would have brought on confusion, forgetfulness, and unpredictable emotions. There would have been a loss of motor skills, an inability to speak, and in the days or weeks leading up to death, partial paralysis and severe dementia. The effects of cerebral atrophy for the men on the Congo-Océan would have been very similar to the condition regularly referred to as AIDS dementia.

Revisiting Pales's autopsies decades later, Pepin, an expert on infectious diseases, found "Mayombe cachexia" to be "certainly suggestive of AIDS." Pales's own analysis and diagnoses ruled out the most likely known culprits: dysentery, malnutrition, tuberculosis, and even cancer. The symptoms of the deceased—generalized lymphadenopathy, chronic diarrhea, and brain atrophy—are all common in patients with HIV or AIDS. Pepin further argues that the work camps of the Mayombe, with their poor sanitary conditions, close quarters, and high male-to-female ratio where some women had more than one sexual partner made it an ideal environment for spreading the virus. One caveat Pepin addresses is the speed with which the disease, if AIDS, killed its victims. While today HIV infection can take a decade or more to become AIDS, the workers on the Congo-Océan would have only been present for a year or two. But "this does not exclude anything," Pepin argues. "For complex virological reasons, it is possible that this incubation period was actually shorter soon after the virus was introduced into human populations."

Pepin's evidence for the existence of AIDS on the Congo-Océan is necessarily speculative and certainly not definitive. By his own admission, "we will never know for sure" if the autopsies conducted by Pales were on men who died from early cases of HIV infection. But Pales's work did provide a credible clinical record of an unknown illness with similar symptoms in a region that molecular biological studies have determined to be near ground zero of HIV-1. Brought out of a secluded corner of Equatorial Africa, HIV-1 plausibly arrived at a worksite in the body of one or more recruits and multiplied and spread. Even if men and women on the Congo-Océan did not die from AIDS, the effort of building the railroad could have helped spread the virus around the region. While HIV-1 did not spread well beyond the region of Pool Malebo until the late 1970s, scientists are quite sure that the disease had been locally present for decades.

With its movement of tens of thousands of recruits from northern Middle Congo and Ubangi-Shari to the south; with its intensification of river traffic from the region that scientists have identified as the point of transmission; with its villagers fleeing across colonial borders; and with its subsequent urbanization of Brazzaville, the Congo-Océan project created ideal conditions for a virus like HIV-1 to prosper. The most likely years of its appearance and early transmission were between 1909 and 1930, roughly coinciding with recruitment and construction. But it is impossible to know definitively if one of the legacies of the Congo-Océan was the transmission of a global disease that has taken tens of millions of lives and counting. For the men and women who suffered from Mayombe cachexia, whatever it was, it surely made little difference. All that likely mattered to them was that they suffered from a debilitating disease in a place most had never asked to visit, many miles from home.

# CHAPTER 9

————

# A BUREAUCRAT'S
# HUMANITARIANISM

THE PREVALENCE OF DYSENTERY, respiratory infections, and undiagnosed cachexia exposed, perhaps more forcefully than other problems, just how unprepared the colony was for the challenges of building a railway line. It was a criticism that was made repeatedly, and often eloquently, not just by journalists passing through or by politicians in Paris but by administrators and doctors on the construction site. As deaths started to mount, the director of the Labor Service was blunt. "It is incontestable that big mistakes have been made," he wrote. The cause? The desire to develop the colony too quickly: "the regional administration lacked prudence, and misunderstood the acquired truth that in tropical countries, for the construction of these great consumers of men that are railroads, one must keep from recruiting labor in regions far from the line." Like many of his colleagues, he also blamed workers' racial inferiority for their ill health: men from distant regions were "too little evolved, too *savage,* let's say the word, to be thus brusquely transplanted." But regardless of the exact cause—lack of adequate medical care or weak African immune systems—the colonial state seemed unable to control whether workers lived or died.

Others were more scientific in their assessments. Dr. Boyé analyzed a one-year period in 1925–26 to consider what caused the annual
"known" mortality rate of 37 percent (he figured that far more died
in reality but went uncounted) and how the colony might go about
minimizing future deaths. The dangers of long-distance recruiting
were statistically undeniable: during that year, fewer than 20 percent of workers from Middle Congo died, while about 50 percent
of recruits from Ubangi-Shari and Chad perished. Part of the reason
for the disparity, he posited, was the fatigue of the voyage from the
north, combined with bad food and insalubrious housing along the
way. But Boyé also found fault with the men themselves: the Saras,
despite their perceived strength and stamina, showed real vulnera-
bilities, such as "pulmonary fragility" as well as the "fragility of this
race to being uprooted." Acclimatization was essential, as men from
the north lacked immunity to diseases endemic to the southern forests. It would have been best, he wrote, to stop recruitment from the
north entirely. But that, he emphasized, was *absolutely impossible.*

Boyé's recommendation was, first, to improve the selection process. It was essential to weed out the vast majority of recruits, select-
ing only the very healthiest. As an example, he cited scrupulous
exams given to one contingent of Saras; over 2,600 recruits were
rejected, and only 900 taken. They were overseen by Europeans,
provided good food including meat, and housed with military discipline. Even with "all the precautions that had been pushed to the
extreme possible limit," some 3.6 percent of the men still died, but it
was a tremendous improvement over the 50 percent norm. The lessons were clear: workers needed better screening, better food, better
housing, better hygiene, and better medical care, including vaccina-
tions. Boyé also echoed other reports that emphasized the need to
keep the men separated by ethnicity, as each race had its "particular
infections" and "pathological patrimony."

In the mid-1920s, however, the colony was simply not prepared to meet Boyé's commonsense recommendations. The number of medical personnel in the colony as a whole had shrunk since the First World War. Funds were available to hire doctors in the wake of the war, but the applicants simply did not come, least of all to Equatorial Africa. A program that allowed doctors to fulfill their military requirement in colonial Africa did not entice a single physician. In 1923 there were only eighteen doctors in all of French Equatorial Africa. In 1928 Chad had a single doctor—that is, one physician serving a possession twice the size of France. That same year, there were twenty-nine doctors in the entire colony, leaving empty half the medical posts the colony could afford.

The construction site had no permanent contingent of doctors until 1925. From that date, there were between three and four physicians on the entire line, caring for numbers that could surpass ten thousand workers, thousands of whom could be sick or injured at any one time. If numbers were against doctors, so was experience. Many physicians who came from France were military doctors, fairly well trained in tropical medicine but inexperienced in dealing with widespread public health issues. They knew little of the local population, rarely spoke African languages, and had to treat men and women who were suspicious of Western medicine and preferred traditional healers. Physicians were joined by a handful of "hygienists"—mainly Russians who'd been trained abroad but were never certified as physicians in France—and a few nurses, both European and African, with limited training. Clinics and hospitals were few in number and far apart along the line; the sick often outpaced the available beds, and the buildings, often little more than huts with earthen or wooden floors, lacked hygiene. There were no laboratories to help with diagnoses or to study illnesses. Little wonder that so many men and women demonstrated so little faith in the colony's medical services.

Illness and death on the Congo-Océan presented multiple and

severe challenges to the colonial government and to Antonetti in particular. Among them was the insurmountable problem of how, with little money and manpower, the colony could practically keep men and women from getting ill and dying. Rouberol and Antonetti expressed remorse about the fallen, but their concerns about mortality were driven less by sympathy than by worry for the consequences to the project as a whole. The sick and the dead slowed the progress made on the construction and increased demands to recruit more men to replace those who could no longer work. And finally the relentless statistics, reported regularly in black and white, drew unwanted attention to the project and jeopardized the continued financial support of the National Assembly in Paris. The word Antonetti received, even from staunch allies of the railroad, was to turn things around.

For Antonetti, the human face of the railroad, the stakes could not have been higher. The uncontrollable loss of life exposed the many contradictions inherent in the arguments he had made for expanding recruitment and pushing ahead with the project in the first place. Historians of imperialism have debated the effectiveness of colonial rule. Some scholars have argued that European colonies were modern states equipped with the administrative and ideological means to reform their societies and transform their landscapes. Other historians, by contrast, doubt that European colonial regimes possessed real power or even ideological coherence. In their view, colonial states struggled even to win over hearts and minds, not to mention transform mentalities or communities. In fundamental ways, Antonetti found himself in a rhetorical and administrative bind between these two assessments of empire.

On the one hand, in promoting the Congo-Océan, he insisted that the colony could undertake a massive project and guaranteed its capacity to safeguard the life and well-being of its subjects—a common standard for what makes a state modern. In speeches and

articles, Antonetti highlighted Equatorial Africa's modern, industrial trappings—an efficient bureaucracy, engineering expertise,
"scientific" knowledge of its human and physical geography, and
the organization of good housing and outfitted medical stations—as
proof of the readiness and efficacy of the colonial government. On
the ground, however, he knew that the colony was sadly lacking
in almost everything. He regularly complained of crushing shortages of administrators, doctors, funds, and basic infrastructure. His
lofty rhetoric of a modern industrial Equatorial Africa financed with
investment capital stood in stark contrast to the forced labor, legions
of porters, and endemic dysentery of the Congo-Océan.

As construction moved forward, then, and as the news of massive
loss of life became public, Antonetti's job as governor-general was
transformed. The experience he had originally brought to the position was undeniable: he had decades of training across France's empire
under his belt, and he had had specific experience recruiting soldiers
and workers as well as administering public works projects. When he
arrived in Brazzaville, the project was little more than an industrial-
scale improvisation, exemplified by former governor-general Victor
Augagneur's decision to start construction before even establishing a
fixed route. Antonetti was expected to bring energy and ideas, as well
as the bureaucratic acumen to order, regulate, and lead.

But the shocking mortality rates reported in the wake of expanded
recruitment made Antonetti's job increasingly one of damage control. Completing the Congo-Océan would certainly require making reforms and improvements to safeguard the lives of workers. But
it would also require vigilantly defending the project from criticism.
To accomplish this, he relied on what he knew best: bureaucracy.
His experience in colonial administration helped him regulate, document, and perhaps most important, *narrate* the construction to its
conclusion. His efforts at regulation brought some vast improvements to those living on the railroad. But those improvements did

not match his accounts of them. He was not content to cover up the losses and miseries created, though he certainly did that. Instead, to guarantee continued and unquestioned support for the project, he fought to transform the Congo-Océan into a humane and humanitarian endeavor.

———————

ANTONETTI'S PERSONALITY SEEMED EMINENTLY suited to the task. By many accounts, what he lacked in raw intellect, he more than made up for with a potent cocktail of authoritarianism, chauvinism, and arrogance. His defenders considered him passionate, possessing "an energy without equal," as one writer put it. His detractors used far more colorful language. Marcel Sauvage, who met Antonetti in Brazzaville in the early 1930s, found him to be "a man of character, in a high position, but of an uncertain morality, a sly intelligence, a heart hardened by avarice and vulgar sensuality." Driven by "pathological vanity," Antonetti had, in Sauvage's mind, "ruined, in ten years, French Equatorial Africa" in his pursuit of economic growth. Albert Londres trod more thoughtfully, giving Antonetti his due, calling him "quick to make decisions, stimulated and not overwhelmed by the size of a task." Londres was no fan, but he acknowledged Antonetti came to his job with a résumé of successes.

The new assignment would have been daunting for even a seasoned political operative. Antonetti, new to the rank of governor-general, faced a steep learning curve. On the one hand, he faced practical challenges, such as how to meet recruitment quotas, curtail the use of porters, complete a service road along the rail line to improve distribution of food and materials, and slow the devastating morbidity and mortality rates of workers. On the other, he had to deal with the bad press that the railroad deservedly received and craft his own ways of responding, deflecting, and denying. During his ten years in the governor-general's office, his techniques and strategies evolved. His entire adult life had been that of a bureaucrat, and so bureaucracy became his tool for organizing, understanding, and explaining. It also became his weapon for confronting criticism, belittling his detractors, and rejecting culpability.

Antonetti was far from alone in his appreciation of bureaucracy. In many ways, the modern French state was built on paperwork. The French Third Republic excelled at the art of designing procedures, documenting, and regulating, not only in civil society but even more so in the military and the colonial administration. Marc Bloch, a towering historian of the medieval period, co-founder of the Annales school, and during the Second World War a member of the Resistance who was executed, wrote of his experience with military bureaucracy in the run-up to the Nazi invasion in 1940. "All the staffs on which I ever worked," he lamented, "had, almost to a morbid degree, the passion for 'paper.' Writing had to be very neat. The style of expression had to be in accordance with an inflexible tradition. All files had to be carefully docketed, and all incoming and outgoing correspondence properly registered." More than simply a paper trail, documentation shaped truth: "From such details is built 'order' in the bureaucratic sense."

Bloch faulted the bureaucratic vision for its singular lack of creativity and innovation. Antonetti, by contrast, embraced it. Having entered adulthood copying correspondence in sweaty offices as a colonial scribe, he was clearly a believer in the power of paperwork. The bureaucratic report—the primary literary genre of empire— was purposefully formulaic, dry, and impersonal. But Antonetti transformed it into a persuasive means not only of refuting and undermining but also of redirecting the terms of discussion about the railroad. His extensive knowledge of his officers' reports allowed him to defend his policies and even turn criticisms of his failures into proof of his successes.

Certain structural factors enhanced Antonetti's authority. The local administration in Brazzaville, despite its subordinate position in the hierarchy, wielded significant influence in the Colonial Ministry in Paris. Notoriously short on cash and manpower, the ministry had little ability to police the behavior of its officials in distant places. It had few choices but to trust the reports it received from Antonetti as accurate. With independent colonial inspectors relatively scarce until well into the 1930s, inspections and investigations were carried out by Antonetti's administrators, many of whom lived in a world with little or no oversight. In cases of abuse of power or corruption, such as that of Georges Pacha, it was not uncommon for the very officials accused to be asked to investigate themselves. The Ligue des droits de l'homme pointed out the shortcomings of this system to the ministry in 1933: investigations led "with the collaboration of local officials, associated with the local quarrels" were done "in vain." But with few options, the Colonial Ministry relied on local administrations to be their own keepers.

Until the 1930s, reports mailed between Paris and Brazzaville took about five weeks to reach their destination. A request from Paris for information from a remote part of Equatorial Africa could take months to arrive at its destination. And even finding local administrators to make reports was no small challenge. The dearth of officials was remarkable: at the time of Antonetti's arrival, there were only about 350 white administrators in all of Equatorial Africa. To offer some comparison, there were nearly two hundred more elected deputies in the National Assembly in Paris than administrators in all of the colony. The combination of time, distance, and a shortage of men meant that large swaths of the colony remained effectively an administrative blind spot, both to Antonetti and to officials in Paris. As Raymond Buell, an American expert on African labor, noted while touring central Africa in 1926, "Parliaments are certainly kept in ignorance of what goes on out here."

Antonetti's authority was also enhanced by the fact that he was a rare constant in an ever-shifting administrative world. His decade as governor-general in Equatorial Africa was unprecedented for its length; most of his predecessors stayed in the post for a maximum of four years. Meanwhile, during his tenure, no fewer than *twenty-one* different men held the job of colonial minister in Paris. The constant flux at the ministry gave Antonetti extraordinary leverage, especially over all matters related to the railroad. He knew what laws had been passed, what regulations enacted, what reforms made, what deals agreed upon—and he quoted from them in correspondence, often to the bafflement of his superiors. He understood the construction plan, the terrain being traversed, and the regions where laborers could be forced into recruitment. He was the primary conduit between the administration and the Société de construction des Batignolles. And he used his institutional knowledge to great effect. In the face of criticisms, he drew on his expertise and authority to deflect or reject or rewrite claims being made about the railroad.

If his distance and institutional knowledge helped win him influence in the Colonial Ministry, Antonetti still faced immense scrutiny and pressure to improve the dire situation surrounding the Congo-Océan. The governor-general responded with indefatigability. The sheer quantity of reports, notes, memos, telegraphs, and letters that Antonetti's administration produced about the Congo-Océan was nothing short of staggering. He was more than dogged; the stacks of faded papers he left behind, be they typed, neatly handwritten, or scrawled, were the evidence of his relentlessness. His output wasn't limited to authoring papers; he also freely wrote notes in the margins of others' documents, often dictatorial orders to underlings on how to respond or react. His strategies for dealing with controversy reflected his temperament and training as a bureaucrat, but they also offer insight into how modern states, when pressured, develop techniques to disavow and reinterpret accusations that are too damaging or disturbing to admit.

His first line of defense was always the same: denial, deflection, and counterattack. From the moment of his arrival, his kneejerk response to criticisms had him looking for someone—anyone— other than his administration to blame. For years, he tried to blame the difficulties he encountered on his predecessor, Victor Augagneur. He got some good mileage out of it. Not until 1928 did a colonial minister finally tell him flatly to stop. "I can no longer keep myself from pointing out," the minister wrote, that "Mr. Augagneur never knew losses proportionally comparable to those that followed your arrival." The minister continued, "And even if one accepts your opinion, which stands at odds with statistics, I must remind you that you've been in the colony for five years."

When faced with criticism from journalists, Antonetti not only denied their findings, he attacked their characters and speculated about conspiracies. When René Maran criticized the deplorable conditions caused by overcrowding on the boats that transported recruits from the north, Antonetti quickly pointed out that Maran had himself been condemned for "abuse of a native." The accusation was true, though Maran maintained that he was innocent, his conviction a result of the racial prejudices of colonial justice.

Antonetti was even less polite when condemning Albert Londres's articles in *Le Petit Parisien* and Robert Poulaine's series in *Le Temps*. In 1929, in a long report to the Colonial Ministry ostensibly about the political situation in the north of the colony, Antonetti spent pages railing against the two journalists' "tendentious campaigns," claiming they were serving Belgian efforts to undermine the Congo-Océan. The falsehoods in Poulaine's series, he insisted, were "so coarse" that they could not "be blamed on his ignorance alone." Instead, Antonetti theorized, Poulaine's reportage had to be part of a Belgian plot to move public opinion with stories of brutality and put a stop to the construction of the railroad. Describing journalists' work in almost traitorous terms, he denounced those who were trying to establish "an extremely prejudicial unfavorable state of mind" against the colony.

Ranting served Antonetti's cause in a number of important ways. He undermined his critics' credibility to an extreme by accusing them of their own morally dubious behavior, as in Maran's case, or of conspiring with foreign governments, as with Poulaine. He also curried support by portraying the colony as facing multiple threats from many sides. And of course, by attacking his critics and their allegedly complex personal and political motivations, he distracted his readers in Paris from the real issues troubling the railroad project. By the time they worked through nearly ten pages of discussion of Londres and Poulaine, who could even remember what the actual subject of the memo was?

A clumsy orator, Antonetti excelled at writing reports and correspondence. They were his platform for demonstrating his incomparable knowledge of the Congo-Océan and the colony. His dogmatism seemed to prove that he had a privileged understanding of what was going on in the colony that no traveler, journalist, or even ministerial inspector could challenge. One example—an October 1925 reply to a report of serious failures of recruitment—provides a taste of his approach. In his dense, eighteen-page typed rebuttal, he ceded nothing, took no responsibility, and even chided his superiors in Paris for not being more sensitive to French, as opposed to African, sacrifices. He denied from the outset any use of emotional language or imagery in discussing the Congo-Océan. Appeals to emotion had a place in fiction, he said, but not in the very serious work of administration; level-headedness and reason must rule the day.

Antonetti particularly protested "respectfully but strongly" the report's inclusion of several photographs of workers on the verge of starvation that the colonial minister had drawn his attention to. The images (now unfortunately lost), he said, could easily—and misleadingly—be used to "move feelings." All the African laborers, he pointed out, came from a "malnourished and notoriously sickly race" and currently lived in camp conditions where many suffered from dysentery (though he took no responsibility for that). Photographs did not allow one "to sit in judgment," he insisted, for they were too easily manipulated. As a point of contrast, he added that the Labor Service often sent pictures "of very beautiful teams" of workers that offered a completely different image of the situation, though he did not include any of them in his letter.

Yet Antonetti was not above sentimentality when it served his own case. Indeed, he quickly told the minister that nineteen out of the one hundred European overseers on the railroad had been evacuated since the beginning of the year. "If one had bothered to photograph them at the moment of their departure," he waxed, "one could have certainly found among them subjects stricken with cachexia." Then, a dollop of guilt: "but as it had to do with white men, that didn't interest anyone." Antonetti's "racial shaming" of his compatriots for caring only about Africans became a key strategy in his repertoire of responses. His aim was not just to instill a sense of guilt in his reader; he demarcated emotional communities along racial lines, forcing the ministry to sympathize with white men or Africans but not with both.

Another of Antonetti's strategies was to bury his superiors in facts. A common, if obsessive, habit of his was to respond to all reports critical of his policies in minute detail, point by point. In these instances, he would produce double-columned documents, with the original criticisms on one side and his retorts on the other; or he would have his responses typed in narrow columns which were then glued into the margins of a troubling report that he then resubmitted to the ministry. His marginalia or commentaries were often considerably longer than the points he was refuting.

Antonetti clearly believed that being argumentative and exhaustive in his denials was more effective than relying purely on accuracy or consistency. In one set of responses, for example, he claimed that workers felt they had been "treated fairly" on the construction site, only then to admit that "the *tâcherons* are not recruited among ambassadors" and were prone to violence. He argued that the crippling lack of administrative manpower made it impossible to distinguish between dead workers and those who had deserted; but he also claimed to have proof that every "freed" worker made it home "in perfect health." Then, as if to confound his reader even further, he chided an inspector for not criticizing the devastating impact that porterage had on the local population.

Were workers treated fairly, or were they beaten? Was the administration able to monitor workers' conditions, or was it sadly overtaxed? It is unclear whether posing such incongruities was a conscious strategy to befuddle his superiors or a by-product of the contradictions inherent in rhetorical claims about his administration's efficacy that did not reflect realities on the ground. Either way, the confusion of Antonetti's responses seemed only to serve his broader case that truths were hard to establish. One inspection report he railed against, he concluded, "shows how all these statistics are variable depending on who interprets them."

If the relentlessness and histrionics of Antonetti's denials and counterclaims weren't enough, he had a final means of obfuscation. As a true believer in the sanctity of paper, he regularly cited previous notes and telegrams, laws and decrees, reports and letters, like a priest citing biblical verses. His citations were an assertion of his infallibility and further mystified his readers. Whatever the combined effect of his denials, the ministry rarely rejected them. Ministers occasionally responded with skepticism and pushed the governor-general for better results. But the powers in Paris rarely chose to challenge the colonial administration's way of seeing. Indeed, if there was one thing that Antonetti's correspondence strove for, it was to prove that he knew more than anyone else about the Congo-Océan. On that point, he remained largely unopposed.

————

DENYING, ATTACKING, ASSERTING, AND even baffling were all parts of Antonetti's strategy of deflecting criticism. All of these moves folded into a broader narrative that the governor-general crafted that portrayed his administration as leading the colony out of darkness and into light. The tale that Antonetti spun was heroic from the outset. It included the by-now-familiar themes associated with the railroad: its importance to the region's *mise-en-valeur*—its economic development—opening "an outlet to the huge reservoir of primary materials" like rubber, cotton, and ivory. The Congo-Océan would certainly be difficult, he always conceded; the region's extraordinarily arduous conditions, its climate and dense forests, and the "state of mind of the native labor—its ignorance" would all have to be overcome. But, as he reminded his ministers repeatedly, "everyone in the Congo wants and considers it indispensable."

The story of colonial development included overt statements about protecting and improving African lives. Antonetti was quick to distance himself from violence in Equatorial Africa, suggesting it long predated the Congo-Océan. In 1926 he excused the violence that accompanied his recruitment efforts by blaming the history of concessionary companies that had acted as if the rights of local people didn't exist. As a result, "there reigns in this country," he said, "a kind of inhumanity toward the native that exists nowhere else. This state of mind, which is at the root of all Congolese stories, will be hard to uproot." Fault lay not with Antonetti or his plans but with the metropole, which was in "very large measure responsible" for allowing concessionary companies to rule the land for so long and for allowing "inhumane" behavior to become "so ingrained." Antonetti at once normalized the violence created by the Congo-

Océan and shifted responsibility for it onto past governments for allowing it.

It thus fell to Antonetti to right these historical wrongs. With the narcissism of an autocrat, he explained, "as soon as I arrived, I had to show a brutal will to action to get this country out of its torpor." Along the way, he even had had to order his men expressly to treat the people "humanely." This message also embodied contradictions that would account for future acts of French brutality. To "safeguard the lives" of Africans, Antonetti once told the colony's top officials, the French must "break their pernicious habits and [submit them] to our paternal demands." Without this effort, the races of Equatorial Africa would die off. In a formulation that turned critics' sensibilities on their head, Antonetti insisted, "humanity forces us, *alas!*, to use severity."

Antonetti called his commitment to humanity "my doctrine." His promise to build the railroad efficiently and humanely was entirely enmeshed in the bureaucracy he oversaw. He and his officers viewed the construction of the Congo-Océan through the eyes of the bureaucratic apparatus. The multitude of reports written, read, cited, and filed were a primary means by which modern empires came to see and understand—to make "legible"—the societies and geographies ruled. People like Antonetti believed firmly that administrations could govern because paperwork made visible the contours of the colony. Reports informed regulations, which in turn became important tools for overseeing and guaranteeing the well-being, health, and security of the population.

Antonetti's "doctrine" was a kind of bureaucratic humanitarianism in which colonial development projects, combined with the careful regulation of the treatment of Africans, would assure a future of civilization, humanity, and prosperity. His unwavering faith in bureaucracy allowed him not only to deny outsiders' criticisms; it also helped justify a more metaphysical argument that inhumanity was impossible in a colony committed to improvement and well-being. The very project of *mise-en-valeur*, as well as the notions of "humanity" that accompanied it, were themselves produced by the paperwork generated by him and his men. The reports, regulations, and decrees were more than mere by-products; they embodied the very ideology, concerns, and expectations of the colony that produced them. Antonetti's bureaucracy did not simply fulfill economic goals and humanitarian norms; it actively defined them. The conception of humanity was measured less by real-life improvements in living and working conditions, or even by lower mortality rates, than by the number of regulations drafted and decrees signed.

Antonetti's vision gazed through the lenses of his bureaucracy. It allowed him to argue that inhumane treatment of workers was aberrant, if not simply unthinkable. The bureaucratic humanitarian mind held that recruits were not intimidated, because recruiters were prohibited from using physical coercion. They were not tied together in coffles, because the yoking of recruits or deserters was forbidden. Workers were not overcrowded on barges, because there were guidelines limiting the number of people transported. Workers did not go without food, because there were regulations dictating daily rations. Antonetti repeatedly assured his superiors in Paris that his orders were carried out with the "minute precision of military orders."

While the Labor Service was meant to investigate mistreatment, inspect conditions, and highlight failures, it ultimately served the administrative status quo more than the men and women it was created to protect. Antonetti ordered the director of the Labor Service to submit monthly reports. Here again the governor-general revealed his bureaucratic prejudices: he seemed moved less by the accounts they provided than by "the care" with which they had been produced, "their minutia," and the detailed supporting documentation that accompanied them. When they reported instances of brutality and extraordinary hardships, Antonetti said they enabled "all

the demanded improvements" to be accomplished "immediately"—
by which he meant not real changes but rather the drafting of new
regulations and orders. Equally important, the monthly reports were
evidence, again not of inhumanity, but of "the effort that had been
made," as well as the extent to which agents "were conscious of the
importance I attached" to improving the lives of workers.

The colonial administration was thus the antidote to—not the
cause of—all that ailed the African laborer. Antonetti claimed that
the Labor Service organized the railroad's building sites "in a fashion that can offer a model for construction sites of private industry."
In addition, he issued decrees "bringing very important improvements," such as increased salaries and rations, better camps, more
medical visits, an ambulance corps, and even the distribution of soap.
Taken as a whole, the regulations emanating from Brazzaville were
meant to account for every aspect of Africans' lives. Antonetti wrote
to the minister, "You will see on every line what care I took not only
for the material, but even the moral, well-being of the laborers."

---

WITH HIS RHETORICAL STRATEGIES and his near-spiritual belief in the
effectiveness of his bureaucracy, Antonetti was remarkably successful at denouncing critics and even twisting inspectors' reports. Mortality rates, however, presented a much greater obstacle. The charts
and tables of the dead and dying—even produced as they were with
misleadingly low numbers—demonstrated to officials in Paris with
cold precision Brazzaville's failures.

In 1925, its first year, the widespread recruitment from the north
delivered what Antonetti had promised: men. More than 6,500 new
men arrived on the construction line to clear forest, move earth,
terrace and build. It was a bold step toward meeting the Batignolles'
quotas and rectifying the gross shortage of workers. More men, new
to the region, however, also meant sickness and death. Reports often

broke up the accounting of morbidity and mortality and blurred the numbers of sick, dead, and deserted workers. But, even so, the correspondence that addressed the health of workers could not avoid the obvious: men and women were falling at troubling rates. In 1925 about thirteen hundred of the year's new recruits—that is, about 20 percent—died working the railroad line. The figure represented more than a five-fold increase over the previous year's losses.

The governor-general deflected, defended, and reassured: improvements were being made, and unforeseen glitches were being dealt with. But 1926 brought worse news. Among the new recruits, the overall number of deaths nearly doubled over the previous year to around 2,500. Put differently, about one-third of the men brought to work on the Congo-Océan in 1926 died. That same year saw a damning inspector's report and a damaging article by René Maran in *Journal du Peuple*. These were followed by Gide's revelations about the Pacha Affair and attacks on the project in the National Assembly. And 1927 was bringing worsening reports of death.

Antonetti was recalled to Paris for meetings at the Colonial Ministry. His trip home was interpreted as a dressing-down for the slow progress and high mortality rates that followed his aggressive campaign to expand recruitment. It is unclear the extent to which his career was on the line. But a few months later he returned to Equatorial Africa fired up with renewed intensity, even by his passionate standards. His taste for attacking critics continued unabated. More pronounced was an unapologetic promotion of the alleged successes of his regulations and policies, as well as an overt manipulation of bad news into good, and good news into triumphs.

His tirelessness served him well because real change came, albeit slowly. Both 1927 and 1928 witnessed more deaths than ever, ranging between one-quarter to one-third of the recruits brought. Statistics were accompanied by more human voices, be they in the form of newspaper accounts or letters from officials. "It is profoundly saddening and even humiliating to note that some detachments of Saras," a military inspector wrote to the colonial minister in 1928, suffered "a loss of 83% of their workforce." The "profound causes" of these losses remained the abuse of men, their transport, poor food, and moral degradation caused by being exposed to "indefinite forced labor." "We must not delude ourselves," he concluded, "the current system is indefensible and cannot be maintained."

But the administration pushed forward unabated. It did, however, adopt recommendations made by doctors and inspectors, though not always systematically. It ordered more rigorous medical exams for new recruits to weed out weak men unlikely to survive the journey to the railroad. It vaccinated recruits against ailments common on the railroad line, including intestinal and respiratory infections, though doctors remained uncertain of the vaccines' efficacy. With progress made on the service road, food delivery improved in terms of quantity and regularity. A new policy in 1928 stipulated that men who fell ill on the site were to be relieved of service. Previously, they had been forced to continue working; many did not live to fulfill their contracts. The administration remained hopeful that these changes would lower morbidity and mortality on the railroad.

Antonetti also took direct steps to ensure the delivery of workers in the numbers needed to continue construction. To meet the quota of 8,000 workers, officials aimed to recruit a minimum of 12,000. The excess accounted for the expected attrition due to illness, death, and desertion. Policies of relieving sick men from their work and overshooting quotas meant a need for more bodies. By the late 1920s, to improve recruitment, officials started paying and issuing new clothes to "liberated" workers upon their return to their home villages so that their compatriots would be enticed to volunteer. In an interview from the 1970s, a former worker said that authorities had compelled him to look sharp and praise life on the *chantiers* when he got home. Salaries, equipment, and clothing,

including blankets and clothes for women, were advertised as benefits of joining the workforce. Women were increasingly welcomed.
As one article put it, women helped improve men's morale, "dispelled boredom, and stimulated men's ardor for work." It appears to
have worked. One administrator touring between Bangui and Fort
Archambault in 1928 wrote of the "excellent effect" that the sight of
returning Saras, loaded with clothes, mats, and equipment, had on
the population.

Improvements started to be felt on the Congo-Océan by mid-
1929. Qualitative reports described better hygiene and living conditions in labor camps, more and better-quality food, and improved
morale. New policies aimed at recruiting stronger men and letting
the sick go home lowered mortality by about 50 percent over the
previous year. *Improvement*—a word that Antonetti and his officials
made the keystone of all discussions of the railroad—was a very relative term. While the number of deaths dropped in 1929, they still
remained roughly equivalent to the levels so roundly condemned in
1925. Beginning in 1929, the mortality rates represented comparative improvements over the darkest years of 1927 and 1928 but still
recorded deaths of around 13 percent of those recruited, dipping
beneath 10 percent only in 1932.

Improvements were uneven and inconsistent, with stark regional
variations persisting. Workers continued to find the Mayombe to
be the worst place to live and work; morbidity and mortality rates
remained high. One inspection estimated that 17 percent of workers
on the construction sites of the Mayombe died in 1929 and somewhat
fewer in 1930. But even within the Mayombe, some stretches were
worse than others. Three camps between Kilometers 105 and 107
lost seventy-nine men and two women in a two-month period, possibly due to water contaminated by medical waste. Despite improved
food distribution, beriberi was still a problem into the 1930s. The
Brazzaville portion of the railroad, across the plains and rolling hills

to the west of the capital, witnessed less illness and death, considerably easier transport between parts of the line, better weather, and more consistent distribution of food and supplies. Reports, however, show that physical abuse continued on both sections, coastal and Brazzaville. There were still cases of white overseers cheating workers out of their fair pay.

The improvements were no doubt welcomed, not least by the men and women on the construction sites. A good indication of improved conditions and morale was the increasing number of men who extended their contracts to continue working. In early 1931, on the Brazzaville side, more than fifteen hundred men agreed to remain longer. It was doubly good news for the administration. It meant less of a need to find more recruits to replace them. And it helped lower mortality rates: studies had long shown that if dysentery and respiratory illnesses did not kill workers soon after their arrival, they were more likely to survive the duration of their stay.

In any other twentieth-century construction project undertaken by an ostensibly liberal regime, officials would have found the continued numbers of death to be utterly unacceptable. But Antonetti heralded the improvements as a complete vindication of his policies since his arrival. They were, to his mind, evidence that his approach had been right all along. "I did everything," he announced with his signature lack of humility; his accomplishments included creating the Labor Service, breaking ground on the wharf at Pointe-Noire, improving the profile of the railroad, balancing budgets, getting loans, and recruiting investors. He castigated the "systematic smear campaigns, fueled by the opponents of French works abroad," and the "false reports "of the "ignorant press"; he even dismissed inspectors' criticisms from years earlier as having been misguided and unhelpful.

What had reformed the Congo-Océan, the governor-general insisted, was his administrative acumen. He had reworked a contract with new expectations of the Batignolles; he had regulated improved food and water distribution; he had achieved the suppression of porterage thanks to the road he ordered constructed. True to his tendency to quote other reports, this time he quoted one of his own letters in which he had championed the efficacy of his bureaucracy: the "spirit of achievement" had been fueled by his commitment to "study, resolve, and perpetually revise" all questions related to labor.

The improved working conditions shifted Antonetti's rhetoric for good. His descriptions of the railroad took on almost religious overtones: the Congo-Océan, he told the colonial minister, was "the one true solution" to the colony's problems. At times, he could sound positively utopian, as when he wrote that "our civilizing, placating action" was creating a new society in Equatorial Africa, "based on peace, order, and regular work" among people prone to acts of "violence and rapine." He pointed to new sources of revenue that the railroad would deliver; metals and minerals— copper, cobalt, diamonds, gold—would be the basis of wealth for the colony, as it would "develop" a "multitude of businesses" from food to construction.

The governor-general showed a remarkable talent for revision, and the improvements on the railroad inspired him to reimagine the not-so-distant past. Improvements in hygienic conditions in the camps were accepted as proof that they always had been adequate. The growing number of men extending their contracts was finally proof that it was "volunteers" who were building the Congo-Océan—a point he made time and again. Antonetti was not the only official to revise history. In 1931, during one of Antonetti's trips in Paris, the acting governor-general, Matteo Alfassa, while recounting the history of indigenous labor in the colony, waxed about the treatment recruits had always received "under the attentive and vigilant eyes of doctors." And while Alfassa admitted that disease had felled "the first transplants" to the worksite, there was not "a sole victim today." Of course, in 1931 dozens of men and women were dying every month on the railroad.

Indeed, optimism could transform the dead. Antonetti employed creative accounting to make the number of men and women who had died—at that point, around ten thousand, by his own officials' estimate—seem insignificant. By dividing the number of the dead into the population of working-age men, Antonetti determined that only 0.33 percent of the population of the colony would be lost— being "thus without appreciable influence on the future development of the race." If, with a twist of the pen, he minimized the deaths of ten thousand men and women, he could also argue that their working conditions had never been wanting. "The accepted sacrifices are undoubtedly infinitely painful," he wrote to the minister, showing a rare degree of sympathy. "But we can say loudly that nowhere, under any latitude, have any worksites been better organized, better supervised than those in the Mayombe." And since precautions had been taken, France could claim the moral high ground: what mattered was that "all that is humanly possible to try to reduce [mortality] has been tried."

———

IT IS TEMPTING TO dismiss Antonetti's maneuvers as artful dodging and cynical rationalizations. But his measuring of humanity in terms of regulation and decrees was not out of keeping with approaches in other parts of the French Empire. The administration of French Indochina, to cite one example, went to great lengths to guarantee the "total care" of various categories of colonial subjects. Regulations, rules, and decrees dictated how political prisoners and plantation workers were to be fed, clothed, and housed in sanitary conditions. As in Equatorial Africa, rules in Indochina did not reflect reality. Despite its efforts, the administration in Hanoi failed to control its own low-level guards, foremen, and administrators, who often flouted regulations and mistreated prisoners and laborers. Food served was insufficient and rancid; men and women were overworked and beaten, lacked medical attention, and often lived in squalor.

There is no doubt that Antonetti's reforms saved the lives of perhaps thousands of men and women. But most of the miseries that they faced in the first place, from recruitment to diseases, were caused by the poor planning and administration of the railroad. Despite Antonetti's powerful verbiage, regulations and decrees did not effectively shape conditions on the ground. His efforts veiled the tough truth that the colony wholly lacked the political legitimacy, manpower, and technologies needed to enforce its own policies. Antonetti's officials were often ill trained, poorly educated, and inefficient. The maps drawn, data collected, and incidents reported provided only a vague picture of the space and lives ruled by the colonial state, often corresponding little to what they purported to detail.

The dozens of men and women who died every month on the Congo-Océan were the proof that Antonetti's administration had minimal control over keeping its workers healthy. Since their deaths happened despite extensive regulatory efforts, the colonial state continued to ascribe high mortality to factors outside the local administrative system—to the monumentality of the project, the disease-ridden forests of central Africa, to workers' supposed laziness, and to the alleged weakness of Africans' immune systems. As one prominent defender of the project put it, these sacrifices, like the loss of men in battle, were necessary to build such a monumental project. "There's not a colonial project that hasn't required burnt offerings."

Despite its disconnection with actual circumstances, the defense of the oversight of the Congo-Océan as humane was remarkably effective. It filled the pages of thousands of reports Brazzaville exchanged with Paris and spilled over into colonial publications. Drawing on this vast body of paperwork, the Colonial Ministry

repeatedly rejected claims of brutality and misery and pushed forward with the railroad, successfully arguing for increases in loans
and budgets to pay for its skyrocketing cost. In a vacuum of established ideals—where workers had no guaranteed rights and where
the very notion of "humanity" lacked meaning—the Colonial Ministry asked only whether the administration had taken every organizational and regulatory step needed to ensure workers' fair treatment. If it had, then the nation had done its moral due diligence.

Such a measure of humanity might not have been terribly humane, but it was effective in the world of colonial politics
during the interwar years. The Colonial Ministry's willingness to
accept this line of thinking, even with its glaring inconsistencies
and detachment from reality, helps explain how a project deemed a
great gift to central African society could also be so murderous, how
a project imagined as progressive and humane could be infected by
such inhumanity.

CHAPTER 10

# SILENCING CRITICS

We tell ourselves stories in order to live.

JOAN DIDION

THOUSANDS OF PAGES OF articles, reports, speeches, and depositions had demonstrated in often vivid detail that the building of
the Congo-Océan brought suffering, varied and unimaginable, to
the people of French Equatorial Africa. Subjects ranging from the
poor living conditions to sordid tales of individual cruelty were discussed in the small offices of the Labor Service and on the floor of
the French Chamber of Deputies, and from foreign ministries in
Paris, London, Brussels, and Berlin to meetings of the International
Labor Organization in Geneva. Evidence and testimonies came
from myriad sources: engaged writers, shocked travelers, politicians,
administrators, and of course the men and women who worked on
the railroad. That the Congo-Océan was a violent and inhumane
undertaking was manifestly clear to many.

The timing of this outpouring of opinion was not coincidental.
In the wake of the Great War, men and women created myriad organizations to improve the plights of communities in need, from refu-

gees and the famine-stricken to children and industrial workers. The campaign to expose colonial violence was part of what George Orwell rather snidely called the "half-baked antinomian opinions" popular after the war, along with "pacifism, internationalism, humanitarianism of all kinds." Many in the 1920s, inspired by the fear and hope that had led to the creation of international bodies like the League of Nations, aspired to forge a new era of peace and humanity.

During the interwar years, new international groups researched and exposed the excesses of colonialism, giving particular importance to abusive working conditions. No Frenchman was more involved in the international effort to reform European empires than Félicien Challaye, who in 1905 had accompanied Pierre Savorgnan de Brazza during his investigations into the atrocities of Toqué and Gaud. After the First World War, Challaye kept his day job as a high school teacher but also went on tours around the world, investigating rights abuses, lecturing about pacifism, and writing about the shortcomings of colonialism. He joined the League Against Colonial Oppression, an organization with a cosmopolitan membership that included British MP George Lansbury, Albert Einstein, and Mrs. Sun Yat-sen. It held its first congress in Brussels in 1927, drawing representatives from over thirty countries and colonies, including China, Cuba, the United States, French Indochina, and India.

Challaye's efforts inevitably brought him into contact with the League of Nations and the International Labor Organization (ILO) in Geneva, which became the center of many new networks of international organizations. Officials at both the league and the ILO were sympathetic to issues of colonial violence, but their effectiveness was limited, in large part because member states curtailed the organizations' ability to comment, let alone legislate, on issues pertaining directly to colonial matters. Internationalists devised new ways to engage with global issues related to colonialism; European empires planned how best to block these organizations' influence and impact. Nonetheless the optimism of the 1920s drove a range of men and women to speak out against the brutality of empire in Geneva and beyond, be it in letters to national governments, in newspapers and books, or at public lectures and conferences.

The debates around the Congo-Océan unfolded in the same spirit. For most critics, the aim of exposing conditions was to raise awareness to bring fundamental changes to how the railroad was using labor. Albert Londres, who as a journalist had uncovered many scandals and miscarriages of justice, dared, as he put it, "to look over the screen" and show what was really happening in the colony. If the French paid more attention to events in the empire, he hoped, then perhaps colonial officials, "feeling the gaze of their country, would wake, once and for all, from such guilty sleep." It was an aspiration shared by others. The ethnologist Marcel Homet prefaced his account of the "suffering" Congo by insisting he had no intention to "heap discredit upon colonial methods"; instead, he aimed to give a "sufficiently human" account in order to "remedy this pitiful state of things." The belief that reporting injustice could improve the lives of distant, often invisible, people was part of the zeitgeist.

Fitting, too, was that many of the critics of Equatorial Africa invoked *humanity* as a reason to act, as a lofty goal to aspire to. The journalist Marcel Sauvage believed that "a little humanity" was what France needed most. When René Maran sent the colonial minister a copy of his *Journal du peuple* article about brutal recruitment conditions, he did so in an effort to promote what he called "a humane and rational colonization" of Equatorial Africa. In 1927 even the former governor-general Victor Augagneur entered the fray, expressing grave concern about the high levels of worker mortality that were being reported by an anonymous source. He implored the Colonial Ministry to launch conclusive investigations into the causes, "to ensure respect for the most elementary principles of humanity" and to protect the future of the colony.

Others turned specifically to the memory of Savorgnan de Brazza to invoke the nation's lost commitment to humanity. Pierre Contet reflected on the previous fifty years of colonization that had achieved so little: "To think that de Brazza lived for this work, that of the humanist, and that he died for it." He continued, darkly, "to think that we established in the Congo a slavery, viler and more tenacious than formerly, and that we built this horror on the principles of 1789!" Londres was similarly disheartened. As he wandered around Brazzaville, still processing what he'd seen on the railroad, he went looking for the humanity that Brazza had represented. "Brazza wasn't there. France has forgotten him."

Humanity, as a distinctly abstract moral motivation, was the domain not simply of professional writers. The lack of humanity—an ultimate condemnation of colonialism's failures—seeped deeply into reports produced by men who worked within the colonial system. A notable example was the colonial inspector Jean-Noël-Paul Pégourier who toured the colony in 1926. Dr. Pégourier did not have a particularly notable career, but he was known in colonial circles for his "doctrinal firmness and his rigid honesty." His judgment was unwavering: the effort to pierce the Mayombe was "inhumane." *Inhumanity* was a term Pégourier invoked in his report's conclusion no fewer than three times, often thoughtfully. Admitting that "the notion of humanity" was "essentially subjective," he nonetheless doubted that the French public would support continuing the railroad if it knew of the "deplorable conditions" discussed in his report. He even compared the conditions created by the Congo-Océan to the "bad treatment" of indigenous people that had helped justify the League of Nations' confiscation of German colonies after the First World War. That "bad treatment" included the concentration camps and genocidal wars against the Herrero and Nama populations in German Southwest Africa (modern-day Namibia).

Roundly denounced by critics as unnecessary, at odds with French ideals, and inhumane, construction of the Congo-Océan still rolled on. For thirteen years, it ground forward, mile upon deliberate mile, never slowing for reassessment, never pausing to rethink recruitment or to build better housing or to import more doctors. The years 1924 to 1928 in particular witnessed an extraordinary spike in human suffering in Equatorial Africa. That Antonetti, his administration, the Colonial Ministry, and many politicians in France knew about it at the time is without doubt. The question that must be answered, then, is how and why—morally, politically, and practically—the colonial administration survived campaigns by respected writers, excoriating critiques in the National Assembly, and international disapproval. The Congo-Océan was a scandal that failed to scandalize—at least to the point where its miseries and injustices were publicly admitted and addressed. Antonetti's brand of bureaucratic humanitarianism, which helped him justify to the Colonial Ministry the effectiveness of his reforms and regulatory precautions, also transformed how defenders of the Congo-Océan came to champion it publicly.

Scholars of human rights and humanitarianism have written histories that show that when the righteous expose wrongdoing and misery, reform and correction of inhumanity follow. This is an inspiring and important history. Tragically, such an optimistic outcome rarely emerged from Europe's empires. White men were all too infrequently the saviors of those they colonized. They certainly produced damning reports, exposés, and public condemnations; but they were met with concerted efforts to deny, undermine, distort, or ignore the truths on the ground. Officials, lobbyists, and politicians deflected accusations of atrocity and abuse; while that story is less inspiring and triumphalist, understanding it is an essential dimension of the history of modern liberal politics, including humanitarianism and imperialism. The Congo-Océan provides an all-too-useful case in point for how the language of humanity could be invoked to explain the deaths of thousands.

———

ANTONETTI WAS THE MOST tireless proponent of his "doctrine"—
that is, his allegedly humanitarian way of reforming workers' lives
through the workings of his bureaucracy. But the spirit and methods
of his doctrine came to inform the opinions of an array of Congo-
Océan apologists. The building of the railroad enjoyed optimistic
coverage in the colonial and even the mainstream press. Many writers championed Antonetti, referring to his good faith and strong
will. In 1926 Camille Guy, a former governor of Senegal, expressed
confidence in Antonetti's approach to regulation and oversight.
Equatorial Africa, he wrote, suffered from a lack of infrastructure,
labor, hygiene, and money; but "on each of these" Guy reassured his
reader, Antonetti "has ideas that seem wise and practical."

In another article, a French senator, Charles Debierre, aimed
to refute claims that the colony had witnessed massive losses of
lives since the onset of colonial rule—a tough argument to make,
especially in 1926. His argument hinged not on facts but on regulations: the improvements needed to strengthen the population were currently being pursued. "Transit routes, hygiene, food
crops, residences, isn't this the entire program of Governor-General
Antonetti?" Even articles that presented the troubles on the construction site in more dispiriting terms—disease, lack of hygiene,
food distribution challenges—expressed faith in the regulations
Antonetti adopted to tackle them.

Elsewhere, Antonetti's alleged accomplishments were relished
in even grander style. Having listed the "inexhaustible quantities" of
goods, from sesame oil to diamonds, that the Congo-Océan would
deliver, the journalist and Africa expert Anselme Laurence assessed
the construction in the most hyperbolic terms. The "tenacity" of
governors-general like Antonetti and their "faith in the destinies
of the colony" drove forward projects that "completely transformed

the economic life of an immense region." Before the Congo-Océan, he wrote, Equatorial Africa had been "nothing or almost nothing. With the Congo-Océan, and Pointe-Noire, it will be a colony that holds an honorable place among all those that we possess." Antonetti couldn't have written better press himself, especially in 1926, when nearly a third of recruits were dying on the construction sites.

In pro-colonial publications, coverage of the experiences of workers differed starkly from the reportage of Poulaine, Londres, Maran, Contet, and others in the popular press. Defenders of empire had their own lineup of prominent writers, many of whom were politicians, current or former colonial officials, and self-proclaimed experts on Africa. Instead of documenting firsthand the abuses and deprivations of the railroad project, or publishing exposés of workers' conditions, the colonial press ran broader discussions of the assumed backwardness of Equatorial African societies. In so doing, it regularly drew readers far afield from the tribulations that faced workers on the construction sites of the Mayombe, to report on less fraught examples of French magnanimity. Campaigns to fight sleeping sickness, a debilitating and deadly disease that killed thousands in the north of the colony, and to expand education were favorite diversions.

For those familiar with the difficult conditions on the Congo-Océan, some articles could be jarring. In 1926, for example, at the very moment when workers were suffering from severe malnutrition due to inadequate food supplies, the prominent colonial commentator Georges Boussenot wrote an article on the "food problem" in the colony. But Boussenot's "problem" was unrelated to the railroad; rather, he wrote at length about what steps must be taken to force African men and women to produce food. "The native does not eat his fill because he does not do any work," Boussenot explained. Entitled, with unintentional irony, "The Congo-Océan Is Necessary to the Life of French Equatorial Africa," the article claimed it was Africans' lack of "foresight" that endangered their future. The state's lack of foresight on how to supply the construction sites—which contributed to the deaths of thousands—went conspicuously unmentioned.

The publication, between late 1928 and early 1929, of Albert Londres and Robert Poulaine's damning exposés made the Congo-Océan the order of the day. Writers in the colonial press—reproducing almost verbatim Antonetti's tactics, such as questioning their commitment to the nation and dismissing the reports as partisan—set out to discredit them. Some pro-colonial commentators simply rejected the accounts out of hand as "lies and calumnies" proffered by travelers who knew nothing of the real lives of people in the colonies. Less aggressive language did the job just as well: many writers referred to the "legends" surrounding the railroad—a clear reference to critical accounts—emphasizing that they "have no relation to reality." Setting the record straight was the goal: a number of the articles and books that refuted Londres, Poulaine, and others even used the same title: *The Truth About the Congo-Océan*.

Some commentators addressed the rumors swirling and faced them head-on. Making light of criticism that the colony was a "vast necropolis of blacks" and that Antonetti was "the great gravedigger," the deputy Louis Proust claimed to have word from a friend that it was nothing of the sort. He mocked Londres's claim that the Congo-Océan was a great drama: *"Eh bien!"* The only drama, Proust wrote, was to be found in the editorial offices of Parisian newspapers. Having cheered on Antonetti—"experienced, courageous, having love for his country"—Proust turned to the truths at hand, which sounded remarkably like those Antonetti told his superiors. And indeed, Proust's information came from "diverse documents" he'd obtained, one assumes from the colonial administration itself. From these, he knew with "certainty" that "the natives . . . eat well, sleep well, and are perfectly cared for." Indeed, his sumptuous description of workers' diets may well have given some of his readers cravings.

More balanced minds acknowledged mistakes and losses, but justified them by pointing to the greater good of development. "In Africa, we must innovate," wrote Julien Maigret, a prolific writer on African subjects who had actually spoken at the Congo-Océan's groundbreaking ceremony in 1921. "We don't have the choice between a good solution and a bad, but between two bad solutions. That's what has to be understood." For the good of Africa, and for the success of France as a colonial power, it was essential to continue with the railroad and not be moved by the "hypocritical clamoring" in the press and coming from abroad. The lives lost, Maigret asserted, would not be "in vain."

Indeed, a number of defenders increasingly—again, like Antonetti—conceded the high loss of workers, asserting it was unavoidable. Gaston Muraz, who'd worked on the Congo-Océan, wrote a short book largely to refute Londres in which he admitted it was a "gigantic work that still kills blacks, yes, but what undertaking of this kind, on such inhuman earth, would be exempt from painful ransom?" Versed in the administration's way of understanding the undertaking, Muraz's book focused overwhelmingly on the improvements brought to the construction site by regulations, rations, and medical attention. Londres's criticisms, then, became old news, as writers, like the colonial administration, shifted the narrative focus from past mistakes to the progress made at the end of the 1920s, when mortality rates began to decline. According to Muraz, the "profound reforms" undertaken by Antonetti and his officials had rendered Londres's account fiction.

For politicians outraged by the horrors reported from the railroad, the writings of Gide, Londres, and others became the basis of the arguments they made in the National Assembly. In 1927 and 1929 Gide and Londres's texts were even cited and read aloud on the floor, placing direct pressure on the colonial ministers who were present in the sessions to respond. But critics were rarely successful at getting anywhere in the debates that ensued. Their criticisms were often rejected in ways that would have made Antonetti proud: facts were disputed, statistics questioned, interpretations challenged. In 1927, for example, Henry Fontanier, a socialist deputy, read Gide's account of Georges Pacha's mistreatment of villagers in his region. The colonial minister, Léon Perrier, dismissed the story as irrelevant to the rest of the colony. "You cannot generalize by speaking of a particular case," Perrier grunted, unmoved. When Fontanier responded that he wasn't generalizing, rather he was stating a fact, the colonial minister replied, "One swallow doesn't make it springtime."

During debates in June 1929, in the wake of Londres and Poulaine's articles, critics effectively pulled apart Antonetti's entire claim that regulations had wrought real change and effective reforms had undone past errors. The colonial minister's best defense—taking a page from Antonetti's playbook—was to refuse to accept any of their purported facts, even when official reports corroborated them. Minister André Maginot, whose name would be notoriously associated with the defensive line of fortified outposts along the German border, neither verified nor denied statistics that his critics quoted. Throughout the debate, he held the journalists' findings, along with Gide's, to be unreliable. "When one wants to know the truth," he told to his critics, "it isn't enough to look for it in a book."

The tactics of colonial ministers, which were entirely in keeping with Antonetti's artful dodging and denying, protected the project from any real danger. Many of the most eloquent critics in the National Assembly were socialist members of the French Section of the Workers' International (Section française de l'Internationale ouvrière), a minority party in the late 1920s. Though not openly opposed to the empire, socialists did take the opportunity to make general proclamations about colonialism's penchant for injustice. Georges Nouelle, a socialist from Sâone-et-Loire, lamented that the public had understandably come to see that "the word 'colony' is inseparable from the word 'scandal.'" He bashed the contracts that allowed the Batignolles to show more interest in profits than in mechanization or in the well-being of their workers. And he denounced the forced labor at the heart of the Congo-Océan: "Forced labor is disguised slavery."

But outraged politicians stopped well short of suggesting that the Congo-Océan should be stopped. Instead, ironically, they called for what Antonetti claimed to be implementing: administrative reforms. Fontanier, who read passionately from Gide's accounts of brutality, spoke for many when he said, "Let me be understood: it's not about suspending work, but to make sure it is done in better conditions." His words were moving; he did not accept those who said "we must build the railway and too bad if it kills lots of natives." But he was far from standing in the way of progress. Fontanier even admitted that he'd met Antonetti and was "struck by his sense of sincerity." In the end, he asked the colonial minister if he could share the findings of future inspections with the chamber. Such requests to revisit at a later date were commonplace in discussions of the Congo-Océan.

In the end, politics ruled the day. During the interwar years, economic recovery from the First World War and the constant economic challenges of the Versailles peace agreement inspired many to look to the empire with pride, though certainly mixed with ambivalence about its economic and political costs. Speeches from many political corners expressed hope that France's colonies could bring economic growth and stability and national grandeur. Few called for major changes to colonial rule. Communists were among the very few in the National Assembly who actively attacked colonial policies; and they were commonly dismissed out of hand for wanting, as one colonial minister put it, to see colonial populations rise up against France. The miseries suffered by the men and women on the Congo-Océan were alarming, enraging, and reprehensible. But righteous indignation was not enough to stop the train of economic development.

———

IN THE MIDDLE OF the construction of the Congo-Océan, internationalist organizations in Geneva took up a question that should have been extraordinarily relevant to the men and women working on the construction sites of the Mayombe: forced labor. Article 23 of the covenant of the League of Nations—"to secure and maintain fair and humane conditions of labor for all men, women, and children"—required the organization to investigate not only slavery but compulsory labor as well. With the 1919 Treaty of Saint-Germaine-en-Laye, the league was committed to investigate "slavery in all its forms" regardless of geography or sovereignty. When the 1925 convention on slavery threatened to include empires, Britain and France resisted. So in 1926, with the signing of an antislavery convention, the league resolved to draft a separate convention on forced labor, including in Europe's overseas possessions. Rather than handling the issue itself, however, the league asked the International Labor Organization (ILO) to direct the process.

In the years leading up to the drafting of the forced labor convention, the ILO had become *the* major international center for documenting and studying labor abuses in the colonial world. It monitored newspapers, contacted travelers for accounts of understudied regions, and collected complaints from indigenous people who had experienced or witnessed abuse. It sponsored conferences and lecture series that included W.E.B. Du Bois, Leo Frobenius, Jawaharlal Nehru, and other prominent figures of the day. And it provided research material for scholars and even advice about publishing books and articles. Officials like Harold Grimshaw, a British lawyer who directed the Native Labor Division, and Albert Thomas, the French socialist who directed the ILO for its first thirteen years,

corresponded constantly, sharing reports, scrutinizing their method-
ologies, and debating their reliability.

The process of working on a forced labor convention was, politically at least, no simple task. Since at least the nineteenth century,
European powers had insisted that the governing of colonies was
strictly a domestic concern and was not open to international oversight. For the convention to have any meaning, ILO leaders had
to persuade European empires not only to discuss publicly their
colonial policies but also to accept how best to reform their practices. Because of the desire not to ostracize important nations in the
league, including the two largest interwar imperial powers, Great
Britain and France, the process had to defend colonial workers'
rights to humane treatment without raising ethical questions about
the legitimacy of colonialism.

Like many colonial lobbyists in France, Antonetti was deeply
skeptical of reformers in Geneva. He viewed, as did many, the efforts
made by the League of Nations and the ILO as attempts to undermine the foreign policy of sovereign states and "to take in hand
the control of our Colonies." He was also deeply skeptical of turning over too much statistical information to the ILO considering
that the interpretation of such data could be manipulated and mis-
construed. But in fact, much of Antonetti's defense of the Congo-
Océan, despite very different interests vis-à-vis indigenous workers,
unexpectedly resonated with the ILO's approach.

It might seem surprising that the Congo-Océan did not attract
significant attention at the ILO. But the organization showed itself
to be very hesitant to criticize member states, especially important
ones. It was in no way anticolonial; like the League of Nations, it
defended the idea of trusteeship and regularly argued that colonialism was the most effective way to bring progress to many parts of
the world. Antonetti's assertion that the Congo-Océan was essential
to the economic development of the region, and to the develop-

ment of equatorial societies, resonated at the ILO. In 1929, when the
ILO held its annual conference in Geneva to address forced labor,
discussions could seem more like celebrations of imperialism than
critical appraisals. The debate that took place among government
and worker representatives from member nations largely lacked references to mortality, abuse, or any suffering whatsoever.

Many of the speeches made would have found eager support
among the defenders of the Congo-Océan. Some speakers even
chose to emphasize the hardships suffered by *Europeans* in their selfless pursuit of empire. Taking the floor, a Portuguese delegate, for
example, reimagined forced labor as a benevolent tool, saying, "as
soon as any civilized country brings civilization to a country in
a lower stage of advancement, it has a right to require a certain
amount of industry on the part of the peoples which it is benefit-
ing." An Australian delegate was more to the point, saying that in
some instances natives had to be made to work because they were
"improvident and lazy." Comments such as these would certainly
have resonated in Brazzaville.

The conference debates were not without speakers who condemned what one delegate called the "whitewashing" of conditions
in certain colonies. But these speakers were repeatedly reprimanded.
The German workers' adviser criticized the ILO for failing to provide any descriptions of the suffering of forced laborers, including
the 25,000 workers who "died off like flies" building the Congo-
Océan—the one time the railroad was mentioned. As relevant as
the German's statement seems to have been, the session president
repeatedly interrupted him, warning him not to criticize specific
colonies or instances of brutality. When the Belgian workers' delegate complained to the president that the German had a right to
speak, the president responded that the speaker had strayed beyond
the accepted lines of discussion.

The ILO's approach to reforming forced labor reflected the

bureaucratic strategies so commonly found onto the Congo-Océan. Creating a forced labor convention required an advisory board, many of whose members were former officials or experts on colonial matters. They drafted a questionnaire that was sent to governments to inquire about what legislation had been established to regulate labor and which specific policies were moving colonies toward the abolition of obligatory labor. The questionnaire could not have been better suited to the approach of the colonial state in Brazzaville. When the ILO asked what decrees and circulars regulated labor, Antonetti's administration listed more than two dozen. Where many colonies had little legislation regarding labor, French Equatorial Africa could boast myriad regulations of food rations, salaries, recruitment, and living conditions. The questionnaire, a form sent to all member governments, did not request information about the implementation of the regulations, or the effects of policies to curtail the practice. Nor did the ILO seek information about the actual experiences of workers, their diets, living conditions, or contracts.

In the end, the predominant argument to emerge from the conference debate was bland enough to be palatable to all: the ILO stood firmly against forced labor but remained sympathetic to the aims of imperial powers. No one captured this tone better than the French workers' delegate, Léon Jouhaux, who said that "the necessities of civilization require the use of forced labour to raise the native peoples out of their present state." He did not doubt that "these races must be raised; it is a fact that they must be taught to work." But he did condemn the use of forced labor, arguing that it never taught anyone anything other than a disgust for and hatred of all forms of labor.

Jouhaux, and others like him, portrayed the prohibition of forced labor not as a challenge to colonialism but as a defense of it. Forced labor was objectionable not because it was brutal and caused suffering but because it threatened to impede the real progress that colonialism allegedly promoted. Jouhaux's words—essentially a reiteration of the central tenets of the French *mission civilisatrice* and *mise-en-valeur*—was the formula the ILO ultimately adopted to walk the fine line between condemning colonial brutality and defending the legitimacy of Europe's imperial expansion.

In 1930 the ILO governing body adopted a convention that called for the end of forced labor "in all its forms within the shortest possible period," allowing recourse to it only as an "exceptional measure" for public purposes. The thirty-three articles and eighty subsections of the final Forced Labor Convention made not a single mention of how specific men and women in the empire experienced labor-related violence or hardship. Even with wording acceptable to all, European empires were slow to recognize the convention. Britain, France, and Holland had ratified it by the late 1930s. Belgium did not ratify it until 1944; Portugal waited until 1956. But even ratifying the convention did not bring major legal changes in the French Empire. Forced labor was not officially abolished until 1946.

In the archives of the ILO, there is a single thin file on the Congo-Océan that seems to have been compiled in 1932. It contains some internal notes written by officials trying to establish what was known about the construction. One memo noted that Antonetti had admitted in a speech in 1927 that some of the workers were "forced"; but by 1931, he'd said they were entirely volunteer—a patent falsehood that no one in Geneva corroborated. Statistics were harder to come by. One official wrote that some in the press had claimed that seventeen thousand had died, but that the number was "warmly contested"; another memo concurred that the National Assembly had never accepted the figure either.

A few short memos typed on onionskin paper represent as well as any document the way in which Antonetti's doctrine for saving *les misèrables* of colonial Africa came to inform interwar humanitarianism. Immensely bureaucratic organizations like the league and the ILO produced studies, recommendations, and reports all aimed at

improving lives through the drafting of regulations. With the oversight and implementation of these laws well outside their jurisdiction,
officials in Geneva contented themselves with knowing that laws
were in place. Paperwork was more reliable than witness accounts,
depositions, or investigations. Indeed, this common administrative
vision was so striking that during Antonetti's leave in 1927, Acting
Governor-General Joseph-François Reste saw no irony in suggesting
that the Congo-Océan system could be "usefully consulted" by the
ILO for "the regulation of native labor" around the world.

While many in Geneva saw colonial abuse as a scourge, their
means of improving the lot of subject populations was not so very
different from those of the very empires they condemned. The space
between regulation and reality—that space so effectively probed by
Gide, Londres, Poulaine, and Maran, as well as the colonial inspectors and Labor Service officials who toured the Congo-Océan—was
a netherworld that bureaucrats and politicians preferred not to see.
Instead, the humanity sought in Geneva, as in Brazzaville, was to be
found in the infallible aspirations of rules, decrees, and regulations. It
was how bureaucracies could reform the world. And the voices of the
men and women on the Congo-Océan were nowhere to be heard.

CHAPTER 11

———

# THE VICTORY AND
# THE FORGETTING

More than any other nation, France has always shown itself
highly benevolent toward the races that have lately come
in contact with our own. While the Spanish completely
annihilated the Caribs of the Antilles, the English cruelly
rarified the autochthones of New Zealand and made the
last native of Tasmania disappear, [and] the Americans exter-
minated *nine-tenths* of the *red-skinned* Indians and relegated
the rest to reservations where they parked them like curious
animals in a zoo, we, at least, are not those to whom it will
one day be asked: Cain, what did you do to your brother?

ÉMILE VEDEL, "HÉRÉSIES COLONIALES,"
*LE MATIN,* NOVEMBER 15, 1934

As Governor-General Antonetti was busy defending the
humanity of his policies and influencing attitudes in Paris and
beyond, the construction of the Congo-Océan trudged on slowly
but steadily. The governor-general's tireless campaigning was as
essential to its completion as moving earth and laying rail. It safe-

guarded the entire undertaking against criticism and ensured that the much-needed funding to complete the railroad would continue to flow to Equatorial Africa. The grandiose and misleading claims helped bury accounts of the continued illness, mistreatment, and death of workers. And ironically, Antonetti's penchant for denial, distraction, and exaggeration ultimately obscured many of the actual accomplishments of the Europeans and Africans who together completed the link between Brazzaville and Pointe-Noire.

The Congo-Océan railroad was an extraordinary engineering feat by any measure, especially considering how, where, and when it was constructed. Some of the challenges that made the construction so difficult were clearly the result of the colony's own inadequate policies. Workers, for example, were regularly tired, unhealthy, and unmotivated because they were forced laborers living in substandard conditions, offered thin rations and limited medical treatment. The lack of substantial mechanization on the construction site was a matter of choice as much as necessity, especially in the Mayombe, where the Batignolles chose to cut corners by relying on tools powered by sweat rather than by diesel or steam. The decision to start clearing the land not from the ends of the route but from many points along the way, even deep within forbidding forests, was a calculated—and poor—choice made by overly zealous administrators.

Other obstacles, however, were impossible to avoid. Equatorial Africa was located thousands of miles from centers of modernized industry, requiring almost all materials—steel, cement, and tools—to be shipped in. Along certain stretches of the route, the climate was inhospitable and the diseases were deadly. And the mountainous Mayombe, with its precipices and gorges, unstable soil and rock, heavy rain and erosion, represented a formidable adversary to even the most intrepid engineer. The fact that such a massive project was accomplished in the 1930s is remarkable. Indeed, it would be decades before the Republic of Congo, relying on far more sophisticated tools, would complete a paved highway for truck and car traffic through the same region. Today a monument and plaque commemorate the day "the Congo defeated the Mayombe" by completing National Route 1; the inauguration of the new highway was held in December 2011.

Defenders and apologists alike described the railroad in terms infused with the hyperbole of propaganda, but even some of the more grandiloquent imagery was not entirely out of place. A common metaphor invoked to describe the railroad's construction, for example, was warfare. Newspaper articles documenting the construction spoke of struggles, defeats, and triumphs. Considering how much of the train was about digging trenches, fortifying hillsides, beating back jungle, slogging through mud, and surviving the inclemency of Mother Nature, writers might understandably have had the recent memory of the First World War's western front in mind. Even the loss of life fit the analogy. As early as 1927, Pierre Mille, a prolific writer on all things colonial, imagined the reforms needed to improve workers' lives in military terms: he called for an "offensive" to be launched, with sound "tactics," to get the job done. If the battle was going to be waged, he argued, "we must win it." Warfare as an instructive metaphor was slow to fade. In the late 1940s, a pamphlet celebrating the centenary of the Société de construction des Batignolles waxed that in building the Congo-Océan, "the equatorial forest and the Mayombe massif were defeated."

If the building of the Congo-Océan was like a war, it was a protracted series of battles that resulted in a muted victory and uncertain consequences for the region. Setting aside the experiences of workers for a moment, the construction proceeded through a number of distinct phases. Until the late 1920s, efforts focused on clearing forest, digging, and terracing. It is impossible to know just how many thousands of tons of plant life and rock were moved, but the amount of terracing alone was staggering. According to official estimates, more than 350 million cubic feet (10 million cubic meters) of earth were terraced in all—a volume equivalent to about five times the Great Pyramid at Giza. The process of removing brush and dirt, and of cutting the massive terraced swaths through hillside and plateau, took tens of thousands of men the better part of the 1920s to complete.

By 1930, the ends of the line were nearly finished. Partially functional, the stretches leading out of both Pointe-Noire and Brazzaville enabled train cars to be used to service the more difficult interior with equipment and manpower. In the mountains, the focus shifted to leveling and to the laborious task of constructing a platform along the route upon which rail was laid. Key decisions continued to be improvised according to developments in the building process itself. In 1921 many had faulted Victor Augagneur for beginning construction without a defined route, but modifications were still being made years later. Even the gauge of the track, initially established at one meter, changed after the arrival of Antonetti. At that point, it was definitively reset at 1.067 meters, a relatively narrow gauge that corresponded to railroads in the Belgian Congo and South Africa. The decision spoke to Antonetti's faith that the Congo-Océan would eventually be part of an international rail network across the continent.

The change in gauge, however, also forced changes in the route. Engineers had to calculate safe gradients and curvatures in the track, in correspondence to the track gauge, so that trains laden with heavy minerals and other goods could make the trip safely and efficiently. Unexpected topographical and geological conditions, sometimes discovered only after terracing began, also upset original plans. Constant modifications of the route meant that, as late as 1930, a detailed article in the colonial press reported that the projected length of the railroad would be 540 kilometers. Changes made in the last years of construction, including the decision to dig tunnels, shortened the final route by some 30 kilometers.

In the end, the relatively short, 320-mile link from the coast

Tunnel of Kilometer 109, Brazzaville side, September 1932.

to Brazzaville counted 1,200 secondary structures, 14 tunnels, and 92 major viaducts. Despite the term, secondary structures were far from insignificant: they alone required hundreds of thousands of cubic feet of cement and masonry work. The broken landscape of the southern Middle Congo necessitated the construction of myriad walls, bridges, drainage ditches, and aqueducts not only to provide support for the rails but also to guard against the dangers of flooding and landslides. During heavy rains, portions of the route could be engulfed by hundreds of gallons of water per second. Some of the drains built were veritable tunnels, eight to ten feet in diameter, reinforced with concrete, to evacuate torrents of water and mud. Along the route, dozens of buildings, including train stations, were also built. Most were small, in a simple French colonial style, with stucco walls, narrow windows, and high slanting roofs, durable enough that they continue to serve their purpose today.

The nature of the soil, as well as faults in the rock, posed challenges at every turn. Tunnels were susceptible to cave-ins and flooding in the interiors, and to landslides at the openings. One such landslide can be seen in a 1932 photograph of a tunnel 109 kilometers from Brazzaville. The dangers of constructing them notwithstanding, tunnels were required for the train to cross the mountainous interior. The longest tunnel, the seventeen-hundred-meter passage under Mount Bamba, took nearly four years to build, with digging undertaken from both ends.

Of the viaducts, more than a dozen were massive edifices, some rising a hundred feet above rivers, canyons, and dense foliage below. Multistory scaffolds, assembled of wood harvested in the forest, enabled workers to pour the reinforced concrete of the towering arches that supported the rail. Photographs capture the extraordinary size and majesty of some of these edifices, such as the multiarched viaduct close to the tunnel at Kilometer 109. The construction of these viaducts, most of which were built between the late 1920s and the opening of the rail in 1934, was as dangerous as it was difficult. In fact, during a tour of one such viaduct in 1932, Antonetti himself suffered a terrible fall. He was very lucky to land on a portion of the scaffolding many meters beneath where he'd been walking. He was seriously injured, but had he not hit the scaffolding, he would not have survived.

Antonetti's tumble was an omen of things to come. A titanic accomplishment in some ways, the railroad soon revealed the prominent shortcomings built into its very design. It was prone to accidents starting even before it was opened. Since most of the route consisted of a single rail shared by both westbound and eastbound traffic, reliable means of communication along the line were essential but often not available. In cases of lapses, accidents occurred: a train would appear without warning and injure or kill unsuspecting workers or villagers, or trains would collide head-on. In 1933 Mr.

The viaduct at Kilometer 108, September 1932.

Girard, a lead Batignolles engineer, was killed with two others when the engine he was on collided with a service train. The steep inclines and relatively tight curves of the rails also had unfortunate consequences. Trains, especially ones with heavy loads, had to travel very slowly, rarely passing twenty miles an hour. Even then derailments, especially along curves, were unfortunately common, sometimes accompanied by loss of material or lives.

In mid-1934 the Congo-Océan was ready to open. Whether it proved to be safe and efficient or perilous and slow, it could now move goods and people from the great and rich basin of the Congo River at Brazzaville to a port on the Atlantic coast. Decades of relying on porters to make the long journey on foot across the Mayombe, or of bowing to the whims of the Belgians and their railroad, came to an end. The battle that had begun, at least rhetorically, with Pierre

Savorgnan de Brazza a half-century before was finally over. And as with many a war, questions loomed about the construction of the railroad. What were its consequences? What was its price?

─────

MAKING INFLATED PROMISES HAS long been the imperialist's cherished pastime. For years, Brazza, Sarraut, Antonetti, and others had promised that the railroad would transform the economy of Equatorial Africa for both Europe and Africa. In the run-up to its completion, self-assured prognostications were made on all sides. One enthusiastic publication predicted that the Congo-Océan would move 800,000 metric tons of goods from the interior to the coast, serving both French and Belgian colonial interests. Critics, meanwhile, dismissed the entire project as a colossal waste of money, an economic and financial disaster. The pertinent question in these debates was less about the Congo-Océan than about the productivity of French Equatorial Africa.

Perhaps the most insightful, if damning, assessment came from Raymond Susset, a politician seriously engaged with colonial policies, who insisted that the railroad was a failure before it started. Its main problem, he argued, was that its defenders had grossly oversold the amount of goods that the railroad would transport upon completion. Apologists had predicted that the very first year of its use, 100,000 metric tons of minerals and raw goods would be exported, and another 50,000 metric tons would be imported. But such numbers, Susset argued, were disconnected from the realities of trade in Equatorial Africa. In the later years of construction, the amount of goods produced annually for export had been under 12,000 metric tons—a fraction of what defenders were predicting would be moved by the train. The colony had also never imported anywhere near 50,000 metric tons of goods needing transport from Pointe-Noire to Brazzaville. The issue, then, was not how much of the oft-heralded riches of Equatorial Africa the train could deliver to foreign markets. Rather, the hitch was whether the colony had ever produced riches in such amounts in the first place.

The answer, at least in the short run, was no. This was not the fault of the train but rather was due to a lack of goods for export. Growth in production was not impossible, but it would require significant amounts of investment. Antonetti and others long vowed that a railroad would bring new money and entrepreneurial endeavors to the colony's interior. Investors, however, were not enticed by the new railroad. The colony had been in an economic downturn since 1928, even before the global economic crisis of the 1930s. Relative to other parts of the continent, the Congo-Océan did not attract private capital: through the 1930s, French Equatorial Africa received less than 2 percent of all investments in sub-Saharan Africa. In the years that immediately followed the opening of the railroad, the tonnage and value of goods moving through Pointe-Noire—a very rough indication of what the train moved—increased only slightly, with imports far outpacing exports. If the Congo-Océan had been meant to release the great riches of the Congo River basin for global export, it did not fulfill its potential, at least relative to the dreams of its admirers.

In her detailed history of the Société de construction des Batignolles, Anne Burnel argues that the Congo-Océan not only failed to help the colony's economy but effectively contributed to its underdevelopment. The railroad's final price—more than one billion francs (well over US$2 billion in today's money)—was paid mainly to the French companies that helped build it. For the Batignolles, it was a windfall, as they took about a quarter of the total budget; it also brought the company great celebrity, despite its decidedly mixed record on labor issues. Over the years of the construction, in order to hide the real cost from the National Assembly, the Colonial Ministry increasingly folded funds for the railroad into the colony's operating budget. Some of the train's budget did go to workers:

about a tenth of the railroad's total cost went to pay for workers' transport, housing, medical care, and salaries. It was not a negligible amount of money, though the bulk of it went into labor expenses rather than the pockets of workers. And with so much of the colony's budget going for the Congo-Océan, it had little left for the development of education, infrastructure, and economic programs.

If the Congo-Océan did not transform the Cinderella of the French Empire in the ways many of its proponents had promised, the train's longer-term value is harder to measure. The most notable changes in its use came after the Republic of Congo gained independence in 1960, when the amount of material shipped along the rails expanded exponentially. In 1962 a northern extension of the railroad was added to link the main line to Mbinda, on the Gabon border. From there manganese ore mined in southern Gabon was transported to Pointe-Noire for export. By 1970, according to a World Bank report, about 1.5 million metric tons of ore was transported annually.

Lumber production also grew rapidly after independence; the Congo-Océan increased annual lumber shipments from about 10,000 metric tons in 1950 to nearly 650,000 tons in 1970. Nearly forty years after its completion, the Congo-Océan finally lived up to the more optimistic prognostications made at the time of its opening, moving about 700,000 metric tons of goods between Brazzaville and Pointe-Noire. It was ultimately the Congolese, and not the French, who made the most of the railroad, transporting millions of metric tons of goods annually from the 1970s forward.

Ironically, the use of the railroad at levels originally envisioned by French promoters exposed the shortcomings of the railroad's original design. The single line made difficult the movement and switching of cars. The tight curves quickly wore out rails and caused breaks in their welded joints, requiring replacement. The use of concrete, prone to weakness and deterioration, caused structural failures over the years, making delays and accidents all too frequent. New technology, including engines and communication systems, has not substantially improved the situation. Between 1990 and 2019, multiple accidents, due to collisions, derailments, and track failures, caused nearly two hundred deaths and hundreds more injuries.

If building the railroad was expensive, so too has been keeping it functioning, an expense that has run into the hundreds of millions of dollars. In 1984 Congolese president Denis Sassou-Nguesso said the maintenance and improvement of the railway was essential to national economic independence. Considering the quantity of minerals, lumber, and other goods moved along it, he was no doubt right. But the cost of the Congo-Océan has also driven the government to look abroad for technical expertise and investment. Agreements have been forged with American, Chinese, and other national and transnational entities, some of which have expected to receive in return what European companies sought more than a century ago—concessions of Congo's rich natural resources. While these arrangements are certainly different from those of the concessionary companies of the early twentieth century, they suggest that the Congo-Océan has never entirely broken free of its colonial origins.

⸻

WHILE THE RAILROAD'S UTILITY, durability, and value has ultimately proven to be decidedly mixed, when it was finally completed, the colonial administration tried to set the record straight for posterity. On the morning of July 10, 1934, a crowd of administrators, dignitaries, and businessmen representing the Belgian Congo, Portugal, Holland, Switzerland, Britain, and of course France gathered in Pointe-Noire. In pressed suits and uniforms, light-colored dresses and wide hats, these men and women came to celebrate the completion of the Congo-Océan railway. For nearly fifty years, French colonial interests had dreamed of building a railroad linking the upper Congo River to the Atlantic Ocean. It was now time for reflection, revelry, and hope. It was also time to try to shape what would be remembered and what would be forgotten.

The inauguration ceremony was part of several days of festivities. Early the previous morning a number of European dignitaries boarded a Micheline—a buslike train car that ran on rubber tires—to make the trip along the rails from Brazzaville. The small stations along the route were decorated with fresh paint and flower gardens. The train stopped occasionally to let passengers stretch their legs, eat at prepared banquets, and fawn over the engineering marvels of the railway. Once in Pointe-Noire, the visitors enjoyed a late-morning aperitif to the sounds of a concert, a lunch at the Hôtel de France, and a gala dinner and ball aboard a ship. A journalist described the atmosphere as one of "free and friendly cheer."

In attendance were also a hundred African "notables." They were clearly not the target audience of the celebration; a commemorative brochure describing the event listed them last in order of importance. Nonetheless, a number of chiefs had been brought on a separate special train from Brazzaville, a day earlier so as not to interfere with the Europeans' journey. A local newspaper said the trip was "a revelation" for these "primitives" accustomed only to walking along paths in the bush. The chiefs' names were conspicuously left off the lists of invitees to the gala and ball. When mentioned at all at the ceremony, Africans were most often portrayed as the lucky recipients of French intelligence and largesse. The focus of the festivities was on French accomplishments. The white community of central Africa had come together to celebrate the achievements, humanity, and grandeur of empire. The rhetoric of colonial subjects working together with the French for a shared future—the rhetoric that had fueled support for the railroad in the first place—did not rule the day.

Public ceremonies offer governments a chance not only to applaud their own triumphs but also to craft their own histories. This event—a dual celebration of the completion of the railway and the beginning of the expansion of the deepwater port at Pointe-Noire— was no exception. Speeches that day spoke of tests and victories, of monumental constructions and profound sacrifices. No backdrop, it seemed, was more fitting to champion Europe's ability to shed light on African darkness than the Congo-Océan. "The penetration of the Continent by a path of such size," the French president of a commercial association pronounced in the first speech of the day, "will allow the country to be developed quickly and our civilization to blossom freely." The railroad represented the completion of what he called "the European project of emancipation" that would help bring a "boom" to the region. And more transformations were on the way: "Pointe-Noire! Yesterday sand, today town, tomorrow great port."

When Governor-General Antonetti took the stage, he announced that Pierre Laval, the current colonial minister who would later head the collaborationist Vichy regime during the Second World War, regrettably could not attend due to the many challenges at home. Laval instead wrote the governor-general a congratulatory letter for his commitment to completing one of "the great works honoring the French colonizing genius." By 1934, France had become divided by financial scandals, rising anti-Semitism, and xenophobia. In February right-wing groups had demonstrated in Paris, fueling rumors of an attempted coup d'état, and leading to riots in the streets and even fisticuffs in the National Assembly, as politicians on the left and right exchanged blows. In the months that followed, French politics had become radicalized between antiparliamentarians on the extreme right and an increasingly unified left, divisions only complicated by the growing power of the Nazi Party in Germany. Antonetti expressed sympathy for Laval's decision not to attend in strongly racial terms: "separated from our *Patrie* [fatherland] we feel perhaps more the dangers that threaten Europe and this white, one can almost say Mediterranean, civilization that took thirty centuries to build."

Although Antonetti had not yet been relieved of his duties, his speech read like his curtain call. He adopted the tone of a man without doubts in his abilities, who'd accomplished the unimaginable. He placed the railroad within the broadest scope of French and African history. France, in the governor-general's words, had saved Equatorial Africa from its own brutality. He celebrated the white men who had died building the railroad. They would, he waxed, "remind us all of how much French blood we have paid for the right to make a Pax Romana reign in these immense lands, yesterday a somber mass grave—where populations have always lived miserably as victims of deprivation, of internal wars, of cannibalism, of the horror of ritual customs—of looting sultans and slave traders." The people of Equatorial Africa, he insisted, were not run-of-the-mill subjects; rather, they had been little more than savages who lived by the rule "eat or be eaten, pillage or be pillaged," and who believed in mysticism rather than reason. Only French rule had "put an end to this barbarity," he said, adding, "we can be proud of the work we've achieved and paid for dearly."

He celebrated the accomplishments of the railroad—the millions of cubic meters of dirt moved, the tunnels dug—with an eye to denigrating his critics, who had always focused on the "mess—disorder—the impossibility of ever succeeding." He chided those who had, he claimed, manipulated mortality statistics of African workers, "inflating the figures to make them more impressive" without acknowledging that all such constructions cost lives. Antonetti also condemned his critics for refusing to admit that "Europeans themselves have not been spared." In the place of such criticisms, Antonetti reminisced about workers who returned home from the Congo-Océan enchanted by their time there, happy with the money they had made and the good treatment they had received. Indeed, one could not forget the "tens of thousands" who had volunteered and even extended their contracts to work longer. The experience had been for the benefit of the people of Equatorial Africa, not for the wealth to be brought by the railroad. Now the people knew not to fear nature and had learned that it could be tamed by means other than magic.

The railroad's original supporters, Antonetti said, stretching all the way back to Brazza, had believed it would transform Equatorial Africa. He assured his audience that it had already done so. Health, wealth, good transportation, and stability had already arrived. The progress achieved had created a "nuclei of white colonization"; projects like the Congo-Océan and the Pointe-Noire port were transforming not just Africa but the entire Atlantic Ocean, making it into a giant "Mediterranean of white races." Standing at the end of his railroad, as the sun warmed the morning air, Antonetti offered a final, grand vision of a future world conquered by white colonization. Brazza's notions of humanity and Sarraut's ideas about the mutual benefits of colonialism—so key to the continual campaigns to get the railroad built—were left unmentioned.

A representative of the military then read aloud the names of the white men—the "pioneers"—who had died in the struggle to build the Congo-Océan. The names, which began with Brazza's, would adorn streets, parks, squares, and stations in the colony, to be honored by future generations. They were to remind all of how, in a "rough and murderous period," these men had won over Equatorial Africa, "penetrated it, pacified it, organized it, allowed it, in a word, to be developed." As a band played a military hymn, the names were read, not a single African name among them. There was a moment of silence and group contemplation; then the Europeans in the audience raised a glass of champagne to themselves and went on with the scheduled festivities, tours, and banquets.

The inauguration coincided with the publication of a collection of photographs, produced by the colony and "signed" by Antonetti himself, called simply *Le Chemin de fer Congo-Océan*. The handsome large-format, leather-bound photo album had gold embossed lettering. Inside, the collection of photographs chronicled, with minimal supporting text, the transformation of points along the line, from Pointe-Noire through the Mayombe and on to Brazzaville. The book, in a kind of time-lapse sequence of images from the beginning of the construction to the end, followed a familiar narrative of *mise-en-valeur*. In sepia tones, forests give way to construction sites; from apparently nowhere, a railroad emerges, cutting through deep canyons, over towering bridges, through deep tunnels. The huts of 1924 are replaced in later photographs with buildings of cement and iron. Turning the pages at last reveals, toward the end of the book, the appearance of European technological prowess: railroad cars, steam engines, and station buildings.

Photographs are often mistaken for transparent representations of reality. The grand book of the Congo-Océan presented an argument as clear and forceful as the speeches made at the inauguration. It did not document the process that Londres had once memorably described as "slitting the throat of Africa, from Brazzaville to Pointe-Noire." Instead, it illustrated empire at its finest: the progress of French engineering over nature, the transformation of a land of broken huts and goat paths into a colony of architectural monuments and smooth steel rails. The narrative, still often associated with imperialism, is one of modernization and development, of rational resolutions to titanic challenges, of the taming of nature.

The more than one hundred photographs capture some of the most extraordinary engineering feats along the railroad, but hardly any of the African men or women who built the railroad appear. With some focus, small, anonymous figures can be seen in corners or faded into backgrounds. In two images, more visible African men are surrounded by modern technology—a jackhammer and a steam shovel—that anyone with knowledge of the construction would recognize as misleading. But the photographs are mostly sweeping shots of the immensity of the undertaking. When human figures appear front and center, as in a striking image of a massive curve of

A viaduct comprising nine ten-meter arches, Kilometer 140 in the Mayombe, 1934.

the rails at Kilometer 140, the person captured wears a white shirt and hat, the telltale uniform of the white man, be he overseer or engineer.

The *Congo-Océan* album, like the inauguration, told the story that many Europeans, particularly French men and women, wanted to hear. It reassured the reader that colonialism was productive and that it moved parts of the world, even corners that were once insalubrious forests, into an age of cement and steel. It aimed to lull its audience into believing that whatever else might have happened in the invisible reaches of the empire, at least something useful came out of it. It suggested that engineers in tidy offices designed and built the Congo-Océan, along with its stations and outbuildings. And it emphatically denied that 120,000 men and women, living in far-from-modern conditions, in fact built the Congo-Océan. After all the controversy over whether workers were treated respectfully or abominably, the inauguration ceremony and the commemorative photograph book ended the drama of the railroad by removing African men and women entirely. Their names were not read out; their faces were not pictured. Life and death on the Congo-Océan was no longer an issue. All that was left were majestic viaducts and sleek rails twisting through primeval forest—and France had built them.

In France, the curtains soon closed on the drama of the Congo-Océan. The most vocal critics of the project's brutality moved on to issues far from the Congo and, in some cases, from colonial violence altogether. Albert Londres had died two years before the line was finished. On a voyage to France from China, his ship burned and sank in the Gulf of Aden. A few days after the inauguration, Antonetti was recalled to Paris and his retirement was announced. He passed away in 1938, some speculated from long-term complications from head injuries suffered from his fall from the scaffolding on the Congo-Océan. Despite the importance of André Gide's revelations, his interests shifted elsewhere, in part to West Africa but primarily to Communism and the Soviet Union, which he toured in 1936. René Maran did not stop criticizing the misdeeds of empire, and he later described colonial rule in interwar Equatorial Africa as akin to Nazism, but once the Congo-Océan was completed, he saw no reason to continue to criticize it.

The railroad quickly faded from French newspapers as more pressing threats loomed. On the day of its inauguration, July 10, 1934, the front-page headlines of France's major dailies gave an idea of the perils the country faced: *Le Petit Journal* warned of "the dawn of Hitlerism"; the Communist daily *L'Humanité* called for a pact for the "fight against Fascism"; and *Le Temps* discussed a battle against unemployment. Among other pressing news items was the recent purging of the ranks of the Nazi paramilitary. The event, known as the Night of the Long Knives, carried out by Herman Goering, Heinrich Himmler, and others to solidify Hitler's control of the party, did not bode well for those hoping for stability in a continent wracked by economic depression. It was all the more alarming alongside domestic political struggles in France and questions about the unity of left-leaning parties to resist the threats from the extreme right. Such existential crises drowned out reports of the completion of the Congo-Océan, a story that received only a few column inches, often buried deep within the papers.

Just as ceremonies, photo albums, and looming wars drove the experiences of African men and women from the public conscience in France, official archives made sure that future generations would be left speculating about the human toll—what Antonetti had called the "accepted sacrifices"—of the project. For all his pride in the infallibility of his administrators to make legible every aspect of the project, Antonetti's bureaucracy effaced and obscured the number of deaths. Statistics and tables from the construction sites did not follow set guidelines or provide strict definitions for the various categories of workers. The use of terms like *dead, sick, deserted, liberated,* and *reformed* all made for what one inspector called "an incomprehensible totalization of dead and living." The lack of a regular procedure was likely due to a combination of poor accounting methods on the construction site, the mediocre administrative skills of officers on the ground, and a distinct desire to keep the exact statistics of morbidity and mortality vague.

In 1929 an official effort was made to tally the dead up to that point, but it was based on the colonial administration's imperfect statistics. No one had counted deaths before 1924. When officials couldn't find reliable data for a period of time, they simply recorded that no one had died. Equally suspicious, after 1929, officials stopped trying to record global numbers of deaths. At this time, Antonetti and others wrongly insisted that mortality was no longer a problem—so why count the dead? The historian Gilles Sautter, citing the most reliable official reports, estimated that the official number of deaths was likely about 14,100. That figure, Sautter noted, did not account for the sick who chose to leave their camp to die alone; including those, he estimated that some 16,000 men and women had died. Sautter's figure is certainly conservative, almost by his own admission. He did not fully discount estimates by Pierre Mille, Raymond Susset, and others that crept upward of 23,000. Put in different terms, it is likely that over the thirteen years of the Congo-Océan's construction, between 15 and 20 percent of all workers perished.

The total number of deaths caused by the railroad could be higher still. Thousands of men and women deserted the railroad construction. Medical inspectors assumed that a very large portion of them likely succumbed before they got home, a hypothesis supported by accounts like Pierre Contet's. Official mortality rates also did not take into consideration the deaths of many porters. Porters who died in the forest between construction sites were considered deserters, not workers. Local men and women who acted as porters as part of their corvée, or labor tax, were not counted among the dead on the railroad, as they were not technically employees. Equally relevant, the men, women, and children who were killed by recruiters, or who died from the dislocation of fleeing villages to avoid recruitment, or who completed their contracts and went home injured, weakened, ill, or permanently disabled—all are left out of estimates of those who died building the Congo-Océan.

The human cost of the Congo-Océan will never be known with certainty. The lives of thousands of men and women, as well as their families and villages, remain invisible to historians. Of course, those who built the train live on in memory, from the deserts of Chad to the forests of the Mayombe. Their efforts are remembered, as one Congolese official put in the 1980s, in stories shared around the fire in local villages. Others are reminded of its history of pain and achievement when they ride the train today. The railroad itself remains a memorial to the great and deadly project accomplished. The fact that the French administration didn't do more to document the sacrifices of those who built it is one of the most blatant and enduring acts of colonial violence.

# CHAPTER 12

---

# THE VIOLENCE OF EMPIRE

This was a signal: for those people, we were no longer men.

PRIMO LEVI, *THE DROWNED AND THE SAVED*

A FINAL QUESTION REMAINS: why was the construction of the Congo-Océan plagued by such abuse and deprivation? The question is simple enough. But when the forms of mistreatment and misery were so varied—from murderous intimidation of recruits to starvation on the construction site, from humiliating kicks and screams to exposure to deadly diseases poorly treated in understaffed hospitals—the answer cannot be simple. It is often beyond the reach of historians to explain why individual people committed individual acts of violence. Why, for example, did a single altercation evolve from frustration to anger to physical abuse? Each case is unique, colored by factors as varied as circumstances, psychologies, personal histories, and individual power dynamics. But if historical methods often struggle to understand behavior on an interpersonal level, they are better equipped to explore how certain specific conditions and contexts make violence—as well as its collaborators, cruelty, and callousness—more likely than in other circumstances.

An array of troubling phenomena, tied both to the perpetration and justification of violence, needs explanation. To begin, why did white men working for the colonial state, the Batignolles, and in related capacities commit acts of brutality or, at the very least, tolerate the neglect and mistreatment of workers? What conditions made it possible and, in certain instances, *expected* that violence would shape white men's interactions with African communities, recruits, and workers? And why did mistreatment so often devolve into utter dehumanization?

Any explanation of the Congo-Océan must start with the ideas about race—especially the unwavering belief in the biological, cultural, and moral superiority of white men over the equatorial population—that buttressed France's very presence in the region. Across the French Empire in the early twentieth century, assertions of racial superiority infused the hierarchy of social relations in which whites reigned as privileged and powerful. White men and women and colonial subjects were often separated by access to basic rights, written into the laws and policies that governed all aspects of lives, including citizenship, marriage and divorce, taxation, and where one could live. In less concrete terms, personal interactions were also defined by regular performances of power, some quite mundane, as when white men and women, regardless of their social status, addressed colonial subjects in the informal *tu* (you) rather than the accepted formal *vous*, a slight that insulted many. Across the colonial world, how one spoke, the angle of one's gaze, and where one stood or sat in certain circumstances were governed by the unwritten but pervasive rules of empire.

The texture of racial distinctions was specific to the historical, geographical, and political contexts of the colony in question. In French Equatorial Africa, where the French themselves had long considered colonization as incomplete in terms of investment, infrastructure, and European settlement, whites defined themselves against their subjects in remarkably straightforward terms. While there was certainly a range of individual attitudes, many Europeans held Africans to be inferior, subordinate, and different. This was not only true in civilian life; it was evident in official circles as well. In a speech on how to interact with the indigenous population, Antonetti expressed bluntly many whites' attitudes: "one doesn't treat a child like a grown up." Government policies promised not just to improve lives but even to save Africans from themselves. Officials repeatedly complained that equatorial Africans needed to be taught basic hygiene, how to eat, how to dress. Failure to develop the colony, Antonetti asserted, would have been to abandon "this country to its primitive barbarism."

The language of white society and officialdom emphasized difference in many ways. In official reports, due to formal expectations, workers on the Congo-Océan—as well as the rest of the colony's inhabitants—were called *indigènes* (natives) or *noirs* (blacks). In less formal circumstances, whites also referred to workers as *nègres*, a term difficult to translate clearly into English but that had a decidedly derogatory meaning. By the mid-1920s, black intellectuals in France had started to reclaim the term as their own as an act of political defiance. According to one Frenchman in Equatorial Africa, everyone knew that Africans found the term *nègre* to be offensive as it connoted a lack of civilization.

The men and women of Equatorial Africa were referred to as "primitives" and "savages" in conversation and in official reports, correspondence, and speeches. *Savage* was not simply a label; it assumed explanatory power. For Maurice Briault, a Catholic missionary who'd spent much of his career in Equatorial Africa, savagery embodied the challenges Europeans faced in transforming Africa into a modern, "civilized" region. Savagery explained why some "tribes," for example, were "quarrelsome, invading, very indifferent to all progress, and very drawn to polygamy." As the chief administrator of the colony, Antonetti echoed such beliefs. In one of his first major public speeches in Brazzaville, he referred to the local inhabitants as living a "savage life" that drove them to misery, laziness, and death. "These races die," he explained, and it was "their own characters" that were responsible for their demise.

If savagery was a category that defined the intractable difference between the allegedly civilized European and the retrograde African, the details of the savage life were as vivid as they were pejorative. Europeans regularly portrayed the men and women of Equatorial Africa as irrational, fetishistic, superstitious, debauched, and cruel. Europeans constantly accused them of that most repugnant of customs, cannibalism, which had preoccupied Europeans in equatorial Africa since at least the mid-nineteenth century. The fact that reliable evidence of actual cases of cannibalism was scant did not stop Europeans from believing it to be a constant threat. In 1905 a history book that was considered to be "a conscientious study of colonial geography" linked cannibalism with other common racial stereotypes of the African as lazy and ill prepared. Cannibals, for example, did not bother to plant crops. In times of famine, they relied instead on "reserves of fattened children and slaves destined for the butcher."

The image of the equatorial cannibal continued to be invoked well into the period of the Congo-Océan. In 1929 missionary journals referenced the region's cannibalism, reminding readers that groups along the Ubangi River had such a taste for human flesh that they ate not only their enemies but their neighbors who'd died of illness. One priest's book, published nearly a decade after the railroad was completed, reprinted a photograph allegedly of four cannibals caught in flagrante delicto. Lest cannibalism seem a special obsession of missionaries, Governor-General Antonetti himself asserted in a public speech that rebellious groups in the colony "would hunt whites" and consume their hearts when they caught them.

The dehumanization of colonial people as savage, primitive, and cannibalistic often went hand in hand with white acts of violence, which themselves were a regular manifestation of the hierarchy and ideology of colonial rule. The coupling of racist ideas with acts of brutality was common across European empires, not simply in central Africa. In French Indochina, for example, metropolitan travelers were appalled seeing white men and women taunt, slap, and beat people ranging from domestic employees to old men in the street. In 1928 the activist Camille Drevet wrote with horror that her hotel in Phnom Penh found it necessary to hang a sign warning guests that it was prohibited to beat the servants. Traveling in Saigon, Félicien Challaye was struck by the poor treatment of the Vietnamese: "I constantly saw Frenchmen offend, injure, brutalize the *indigène*." Revealingly, he noted that his compatriots hit "for the pleasure, or well, as they say, to maintain the prestige of the white man."

The connection between racial prestige and casual (and not-so-casual) violence was clear in French Equatorial Africa. Robert Poulaine, traveling there in 1928, was struck by the number of white men and women he encountered, be they recent arrivals or veterans of many tours, who spoke of their utter disillusionment with Africans. According to Poulaine, they'd say, "They have all the faults—liars, thieves, lazy, and above all stupid brutes." This attitude was raised to the level of dogma, he noted, leading to an "always contemptuous, sometimes brutal attitude." Derision was the handmaiden of violence.

Poulaine's observations were not news to the men and women of the colony, who had been filing complaints for years. In 1916, for instance, villagers in Lambaréné, Gabon, just to the north of Middle Congo, penned the first of a series of letters denouncing local French officials for abusing them. Their mistreatment was, in their own formulations, inextricable from their race. "The contempt," they protested to a regional inspector, "that the European has for the native means that one is always treated as a '*sale nègre*' [dirty negro]." The villagers spoke of brutality and injustice, including stints of forced labor in which men and women were locked in holds of barges, with no access to bathrooms, and "brutalized." Showing their fluency in French republican rhetoric, the villagers asked, "How do you want us to become 'French' if we are slaves and not the children of France? Where is the fraternity? And where is our liberty?"

Despite multiple reports of humiliation and abuse, as well as the intervention of the Ligue des droits de l'homme in Paris, the administration did little to address the complaints. Villagers persisted: "the *nègre* before the white man," they complained, "is considered a bestial being." Frenchmen acted terrified of Africans, keeping them at a distance, even in public administrative offices, for fear of getting fleas or jiggers. One administrator, they claimed, refused to hold any afternoon meetings because, at that hour, the official said, "it was too hot to talk to blacks." This rude behavior came on top of beatings, injuries, and insults. Showing the extent to which racial hierarchies in the colonies could be internalized, the men demanded only fair treatment: "Our desire is not to be equal to the European, but we desire certain improvements of our lot."

For more than a decade, villagers' complaints from this one region piled up. In 1929 another Gabonese group claimed that their local administrator, a Mr. Gamon, had committed crimes ranging from kidnapping and rape to forcing everyone—even the old, sick, and pregnant—to work in abusive conditions. They showed prescience about the paradoxes of an ostensibly liberal empire: "if Monsieur Gamon doesn't like the natives," they asked poignantly, "why did he accept to serve in a country where everyone is black?" The Ligue des droits de l'homme wondered the same thing, demanding explanations for such examples of mistreatment. Although the villagers' accusations were confirmed by *ligue* members in Libreville, the administration claimed that nothing could be done for lack of "proof."

In their complaints, the villagers pinpointed the problem with their treatment: they were at once, at least rhetorically, men and women living in an allegedly liberal republican empire, and yet they were simultaneously treated as less than human. They identified not a contradiction but rather a key dynamic of colonial power. Physical and emotional abuse fulfilled a dual function: abuse was an interpersonal statement, a quotidian reminder of subordination for the target of mistreatment. It was also a performance of control, an assertion of political, social, and juridical power on the part of the abuser. Europeans knew that most often they could yell, hit, or beat without any ramifications whatsoever. When they felt threatened, fearful, or in need of demonstrating the control they wished to have, they often turned more cruel. Sometimes there were consequences for abuse, but in most instances—including in the cases reported in Gabon— there were none.

Again, racial distinctions were paramount. A racial esprit de corps in the colony made it unlikely that Europeans would denounce one another for the mistreatment of Africans. Those who did were often castigated by their peers or criticized by the administration. This is not to say that all white men and women lived together in harmony. As the criticism of the immoral ways of non-French *tâcherons* on the Congo-Océan made clear, there were definite hierarchies and prejudices among European colonial populations shaped by nationality, religion, and many other distinctions. Other kinds of tensions abounded as well. Catholic missionaries, for example, complained privately to their superiors in France about the ill effects that heavy-handed recruitment and difficult working conditions on the Congo-Océan had on the religious life of converts. But they steadfastly avoided open, public criticism of the project or its mismanagement. Europeans remained largely unified around a collection of shared ideals, goals, and privileges. When behavior turned violent, the white community protected its own, in a troubling colonial perversion of noblesse oblige.

The near immunity of Europeans was not without its detractors. Throughout the first half of the twentieth century, various colonial ministries in France condemned the "disproportionate" sanctions given to Europeans and indigenous people for the same crimes. "Verdicts of race," the ministry warned, risked undermining the principle that "justice is equal for all." Aware of the delicacy of racial hierarchies in the empire, though, the ministry insisted it wanted not to "diminish the prestige of the European" but rather to emphasize that the settler's authority should be defined by his "magnanimity and his moral superiority," not by "his power."

Official inaction in cases of violence, however, remained the norm across Equatorial Africa. Time and again the administration chose to ignore cases of abuse that ranged from humiliation to murder, many of which, as one newspaper put it, were left in the shadows. Journalists, travelers, and colonial subjects all voiced disapproval, warning that it was completely out of step with the French commitment to "civilization." White mechanics could kill a black man without being charged; a high-ranking official could beat his servant "boy" to the point of unconsciousness without fear of retribution. The men and women who lived under French rule were scandalized by the perpetrators' deficit of empathy and lack of accountability. The only explanation for denying justice to the people, according to one journalist, was "negrophobia." While not the sole cause by any means, negrophobia fired the cruelty with which much violence on the Congo-Océan was forged.

———

A STRONG BELIEF IN the racial, intellectual, and cultural inferiority of equatorial societies; a tendency of white men and women to defend one another regarding violence toward Africans; and a tradition of administrative neglect of investigating and punishing the abuse of local people: all these provide essential contexts for understanding the story of the Congo-Océan. Combined they created an environment where many white men considered the lives of workers expendable, and their mistreatment legitimized by a system of administrative oversight that was uninterested in dismantling white brutality.

Clearly the colonial administration tacitly accepted certain forms of violence in most aspects of the construction of the railroad, from recruitment of labor to work on the line. It accepted "light" abuse—much of which was not light at all—as an inevitable part of building the Congo-Océan. Even officials from the Labor Service, expressly created to protect the well-being of workers, accepted yelling, badgering, and slapping as unavoidable. As director, Jean Marchand had this sort of mistreatment in mind when he spoke, with little concern, of white overseers' tendency to be a bit "too rough" on their workers. Violence became an administrative concern only when it endangered the construction effort, usually by driving workers to walk off the job, or when it ran the risk of attracting public outrage, either by seeming to be systemic or by resulting in cases of gratuitous cruelty.

How, then, to account for the prevalence of behaviors that crossed the line from acceptable to excessive? One of the more common explanations points to the effects of the brutalization of Europeans. As a concept used by historians, brutalization is most associated with the experiences of soldiers in modern warfare. The relentless violence and emotional hardship of particularly intense experiences, like combat on the Western Front during the Great War or in the Pacific Theater during World War II, is believed to strip soldiers of the morality, empathy, and emotional stability that guides them in peacetime. The brutalized soldier imagines enemies in ways infused with racial prejudice that effectively dehumanizes them. Confronted with life-or-death situations, and faced with an enemy infused with racial hatred, the brutalized individual becomes prone to acts of extreme and unnecessary violence, even by the standards of modern combat. The results of brutalization can include torture, the wholesale killing of civilians, the defilement of bodies, and other obscene atrocities.

Commentators at the time and scholars since have suggested that a similar transformation could be seen in Europeans who ventured into Africa, Asia, and beyond in the age of empire. Abroad, white men and, to a lesser extent, women could become aggravated and antagonized, weakened by inhospitable climes and fear of people they imagined to be menacing and savage. Under such conditions, the theory goes, they broke psychologically, often erupting in uncontrollable bursts of anger and cruelty. Memoirs, diaries, and letters of adventurers, explorers, and scientists document the tribulation of remote travel that drove some out of their minds. Equatorial Africa posed a particular risk; even in the Mayombe—in the middle of a construction site—white overseers sometimes encountered other Europeans only infrequently when an inspector, doctor, or traveler passed along the worksite.

The idea of the "bad" white man in the empire has a long pedigree. The European driven mad by the tribulations of life in the non-European world can be found in the works of writers of fiction from Thomas de Quincey to Doris Lessing. No fictional character embodies the idea more than Kurtz in Joseph Conrad's novella *Heart of Darkness*. As much as any literary character, Kurtz has come to personify European anxieties about modern imperialism. He represents both the promise and the threat of empire: a brilliant colonial agent of unparalleled success at harvesting and stockpiling ivory who also decorates his compound with the heads of murdered Africans. The world in which Kurtz lives, surrounded by cannibals and dense jungle, is painted as so vividly awful, it convinced the Nigerian writer Chinua Achebe that Conrad was "a thoroughgoing racist."

While Kurtz might be "the twentieth century's most famous literary villain," his legacy has been far from purely fictional. Writers as varied as Hannah Arendt, in her classic critique of imperialism, and Adam Hochschild, in his much more recent account of atrocities in Belgian king Leopold II's Congo Free State, have speculated on who Conrad's "real life" inspirations were. The hunt for the *real* Mr. Kurtz—and the assumption that there was one or more—reveals the reach of Conrad's vision. His fictional character has become the historical ideal of the brutal European abroad and of what could go wrong in Europe's empires. Kurtz's brilliance and cold-bloodedness capture Europe's conflicting impulses: altruism and greed, reason and insanity, humanity and cruelty.

Kurtz-like figures loom large in Arendt's understanding of why modern European empires were prone to outbreaks of violence. African colonies were a place, Arendt argued, where gentlemen and scoundrels rubbed elbows, freed from European society's annoying moral norms. There, they "felt not only the closeness of men who share the same color of skin, but the impact of a world of infinite possibilities for crimes committed in the spirit of play, for the combination of horror and laughter." Race was, for Arendt, a major motivation for brutality. Faced with "a whole continent populated and overpopulated by savages," Europeans were driven to massacre, decimate, and murder to get their way.

Arendt's insights on the lure of adventure and freedom from social and moral norms contain much truth. When polled in the 1960s about their motivations for wanting to live in the empire, a large number of former French colonial officials cited their desire to be liberated from strict social norms as well as the draw of exotic locales. Enticing, too, was the assumed nobility of work in the colonies; images of real-life heroes, including Brazza, as well as literary characters in popular works by Pierre Loti, Rudyard Kipling, and others, convinced young men that they were civilized gifts to humanity.

Men imagined the colonies to be a place of action, where they could lead and where people would follow. One administrator who had served in the 1920s said that a colonial career allowed a white man to be "a chief." In a 1926 novel by the French writer and life-long colonial figure Robert Delavignette, the character of a young administrator reassures himself with a pep talk: "I am chief because I have white skin, straight hair, and a pointed nose, which gives me more than human dignity because here man is black, curly-haired, and flat-nosed." Appearances had absolute moral and legal weight: "For me everything is easy," the administrator tells himself, "everything is permitted." Race brought the alluring fantasy of infallibility. Such imagined power was appealing in the all-too-common cases where the men venturing out to the colonies were not the *crème* of society's *crème*.

The potential dangers of letting overly confident and morally dubious white men loose in inclement lands were not lost on commentators at the time. Georges Hardy gave sustained thought to the impact of the colonial life on men who served abroad. While still in his twenties, Hardy had been an inspector of schools in French West Africa, soon rising to oversee educational reforms there and in Morocco. Along the way, he earned a doctorate from the Sorbonne, writing a thesis on colonial history. In 1926 he was named to direct the École coloniale, France's premier institute for overseas administrators. In 1929 he reflected on the dangers of, essentially, colonial brutalization. The administrator living in tropical humidity that "strains the nerves of the European," he wrote, needed "a solid frame of consciousness to remain master of himself." No one could expect him to be "a saint" or a "model of all virtues," but he had to be expected to avoid "bad habits, such as vulgar language, tantrums, and brutality."

The consequences of a white man's downward spiral were serious: loss of dignity and control on a personal level meant disorder for the colony. According to Hardy, colonial service was a great moral test. Men who joined always underestimated the toll taken by melancholy, homesickness, and the "anguish of nights deep in the bush among forces thought to be hostile." The sum of these challenges caused nothing less than "a peculiar and cruel moral destitution," a state of mind that could result in the most unfortunate behavior. The recruitment of administrators, Hardy warned, needed to target men of "a vigorous morality" and provide them with "moral training" to prepare them for the myriad challenges and temptations they would encounter. Even the most compassionate European soon became frustrated; the young and idealistic "indigenophile" risked becoming a seasoned and jaded "indigenophobe."

The question remains, however, whether brutalization provides a useful explanation of colonial violence. Many of the tribulations described by Conrad, Hardy, and others—the difficult weather, the frustrations of life among the "uncivilized"—were echoed by officials on the Congo-Océan. French officials, as well as employees of the Batignolles, were quick to identify acts of violence with midlevel white overseers who were said to work and live under particularly difficult circumstances. In the dark and dank Mayombe, these men regularly lived alone, lacking the comforts of home. Their inability to communicate with their workers did little to improve prejudices: the futility of trying to explain engineering plans to teams of men who did not speak the same language no doubt frayed nerves. This frustration was compounded by the irritations of insects, humidity, and darkness, all of which ate endlessly at their patience. Some drowned their miseries in alcohol, a pastime that did little to improve their dispositions or outlooks on life.

In the first years of the construction, the Batignolles had hired Frenchmen, as an administrator described them, "of a sure morality, well paid . . . disposed to organizing and caring for the worker." But within a few years, few Frenchmen were enticed by the work. They had to be replaced with less savory "foreigners of every provenance, linked to the company by very tough contracts." *Tâcherons,* paid by the job, were concerned solely with getting the most work out of their laborers as possible, pushing them to work faster and longer hours. The Batignolles showed little interest in the treatment of laborers, offering minimal incentive for European overseers to treat their men with care. With these "second rate" overseers, one official added, violent conflicts were inevitable.

Marcel Sauvage found that workers shared this view of their white bosses. Workers certainly had complaints with the way Frenchmen treated them, he reported, "but they have only hatred—and well-nourished hatred—for the Spanish, the Italians, the Germans," whom they perceived to be "the most maladroit and ferocious colonial overlords." According to Sauvage, the workers were fond of saying, "there's the white, and there's the black, there's the monkey, and after that there's the Portuguese." If accurate, Sauvage's reportage suggested a point upon which African and French opinions overlapped. Workers' apparent disdain for Spaniards, Italians, and Portuguese corresponded to popular prejudices in France about Mediterranean immigrants. All white men were not the same, for the French or for their workers. And yet, of course, the Batignolles still defended these men's right to use physical force to motivate their workers.

Physical and psychological strain certainly did little to minimize violence and cruelty on the Congo-Océan. But while white engineers and overseers endured widespread hardships, a causal link between them and the systemic mistreatment of workers is impossible to qualify. Some white men who lived and worked in extraordinarily tough conditions were never accused of abuse. The white men, for example, who dug the 1.7-kilometer tunnel through Mount Bamba—"this thankless mountain"—in the heart of the Mayombe faced daunting challenges. Working the tunnel was a dangerous and exhausting endeavor, often done in darkness, ankle-deep water, and at risk of cave-ins. One of the few places on the site with mechanical drills and jackhammers, the noise was deafening. Temperatures in the tunnel vacillated between extreme heat and cold.

Assisted by many workers, the white men who oversaw the tunneling were largely Polish, Czech, and Italian—that is, just the non-French workers that the administration considered morally questionable. They were paid on average about 2,000 francs a month, some forty times the salary of an African worker but not considerably more than wages at home. On their salary, these men were expected to provide their own food and shelter on the line, however they might do that. To save cash, many of these men chose to live, according to one report, "like the natives." To make white men exist like this, one report concluded, was to treat them "without humanity"—itself an unintended admission that African workers were treated inhumanely.

If difficult living and working conditions brutalized white men, then it follows that those in the worst circumstances would be the most prone to racialized hate and violence. And yet these men, who oversaw dozens of African workers, were not regularly accused of acts of brutality. Indeed, if anything, European tunnel workers seem to have been more collegial with their African teams than were white men on other worksites. A photograph from around 1933 showed white tunnellers side by side with their African workers. White employees were very rarely photographed with their workers. When they were, it was most often in situations where Europeans, dressed neatly, stood by as their workers sweated. Here, the white men did not even dominate the center of the gathering, as the central white figure has an African co-worker to his left. White and African tunnellers shared not only the same space but the same hats, boots, and clothing as well, lending the image a rare kind of visual egalitarianism. Photographs on the rail-

Workers and overseers at the entrance to the Bamba tunnel, c. 1933.

roads were governed by the segregation of the society; this one bucked the rules.

Another reason for skepticism is that the "brutalized" white man was a favorite excuse provided by officials in Brazzaville. In the midst of the Pacha Affair, for example, Antonetti deflected criticism of the brutality of recruitment by insisting the problem lay instead with a shortage of manpower that made relying on "unbalanced and even criminal agents" a necessity. In a letter to his superiors, he pointed to a number of cases of isolated administrators burning villages and killing indiscriminately. In rich and disturbing detail, he recalled one official who, hoping to elicit a confession, had tortured and killed an African man accused of stealing a gun. Soldiers beat the suspect, put chili in his eyes, knocked his teeth out, and burned his thighs and sexual organs. He died after forty-two hours of being tied to the ground with his arms stretched out, in the heat of the dry season, without food or water. Although the official was clearly out of his senses, Antonetti noted, he remained employed in the colonial service due to the desperate shortage of administrators.

It was an unusual argument for a governor-general to make, especially one trying to assuage his superiors' concerns about violence and public opinion. But Antonetti hoped to convince the Colonial Ministry to send more administrators and agents to ease the burdens faced by men in isolated posts. The presence of more men would allow Antonetti to remove "these most corrupt agents"; failure to do so, he warned, would have "the most dangerous consequences." Young agents who arrived at the colony full of "zeal" soon came to realize that, among certain criminal "black sheep" in the ranks, "murder is an unimportant peccadillo, and that the worst abuses are only a means of command." Without more men, he warned, revelations "of barbaric crimes that outrage our conscience and our feelings of humanity" would continue. The "shame" of such crimes could only hurt "our entire project."

Evidence of such criminality and psychosis actually helped Antonetti *defend* his administration. Pointing to the inhumanity of these men transformed the violence of the entire project into a series of unrelated and isolated episodes. If there was a problem, it was not with the inherent brutality of colonialism but rather with funding from Paris; if only the ministry assigned more staff to provide the manpower needed to rule effectively, then Antonetti could purge criminal behavior. The officer who put chili in a suspect's eyes was but a regrettable exception to an otherwise orderly, if understaffed, project. The problem was not that the colonial administration lacked order but rather that isolated individuals lost control under circumstances that could be reformed and improved.

The belief that rogue villains alone were responsible for the colony's excesses offered an odd kind of comfort. Criminals were something comprehensible, someone to blame. The violence and misery of the Congo-Océan stemmed from problems that reached beyond the actions of unrelated individual "bad" Europeans. Racist beliefs and brutalizing conditions existed in many early twentieth-century European colonies without resulting in the kinds of atrocities and loss of life witnessed on the railroad. The viciousness of the Congo-Océan was not, after all, limited to beatings or torture. It encompassed the intimidation of villages, the chaining of recruits, the starvation of workers, and many other deprivations. These were systemic problems, not simply the deeds of lost and broken individuals.

TO RACIST IDEAS AND trying physical and psychological circumstances must be added the pressures and expectations of trying to build a railroad in a part of the world almost completely lacking infrastructure. Building the Congo-Océan required a broad, efficient system of industrial organization, engineering expertise and technology, and bureaucratic oversight. Equatorial Africa was gravely lacking in all these areas even before Antonetti's arrival. His new plans heaped on more challenges without providing any new solutions. Instead, his major policy shift heralded massive recruitment of tens of thousands of "volunteers" from the distant reaches of the possession. The colony would need new housing, new supply and transit routes, and medical attention for workers. The belief that the administration could overcome these obstacles was colonial hubris at its finest.

The challenges of completing the Congo-Océan, especially in the face of mounting criticism in the press and in the National Assembly, put Antonetti's administration under extraordinary pressure from the outset. The effects were palpable. Antonetti's interaction with Rouberol, the lead engineer at the Batignolles, was cordial and respectful but tense at best. While everyone in a position of authority blamed inclement weather and tricky terrain for the slow progress, Antonetti and Rouberol also eyed one another critically, each blaming the other for the construction's shortcomings. Antonetti blamed the slow advance through the Mayombe on the Batignolles' refusal to use modern, industrial machinery, while Rouberol blamed it on the colonial state's failure to deliver workers. Engineers could mathematically calculate the advancement of construction: fewer men equaled fewer work hours equaled less earth moved. The administration's culpability for delays, caused by not producing enough workers, was literally calculated on the Batignolles' monthly charts.

The competitiveness and mutual resentment that simmered between the governor-general's office and the Batignolles did nothing to promote the workers' health and safety. Rouberol dismissed criticisms of any kind from the governor-general with, at best, strained politesse. In 1926, when Antonetti encouraged the Batignolles to reduce the number of porters used because of the misery the job caused, Rouberol replied tartly that the company's agreement with the government did not forbid porterage, and quite simply, it was necessary. He then turned the tables and criticized Antonetti's failures: "Believe me, Monsieur le Gouverneur-général, that it isn't for our pleasure that we speak so often of a lack of men." The question of labor dominated all others, he said, because it undermined their entire program. Under pressure to build as quickly and cheaply as possible, the Batignolles repeatedly put aside discussion of how it treated workers, or whether they lived or died, in order to harp on the single preoccupation of having more men.

Pressure drove both men to make decisions that cast thousands of workers' lives in danger. First, Rouberol's insistence that the state must deliver thousands of men had the immediate effect of flooding the construction site with people who could not be properly housed or fed. In a rare acknowledgment of guilt, the Colonial Ministry admitted—years later—that this had been a mistake. The eight thousand men that the state agreed contractually to provide, the minister admitted, "exceeded the possibilities of the colony." To recruit that number of men obliged administrators to search hundreds of miles away, exposing them to difficult travels and unfamiliar climates. Multiple officials highlighted the "alarming" conditions caused by the "too hasty recruitment" of workers from the north. Recruiters were not only brutal: they enrolled weak and sick men, who could fill their quotas as effectively as strong and healthy ones. If men and women died farther down the road, it wasn't the recruiter's problem. In one politician's opinion, this practice, a result of Antonetti's lack of preparation, "turned this poor colony upside down."

Distant recruitment exposed the weaknesses of the colonial state and caused French standing across much of the north to deteriorate. Recruitment closely coincided with the Kongo-Wara rebellion, or the War of the Hoe Handles, involving a movement started by Barka Ngainoumbey, a Gbaya spiritual leader who went by the name Karnu (or Karnou). In part, Karnu inspired followers by encouraging them to refuse to work for French concessions and other projects, including the Congo-Océan. Karnu's movement was not exclusively an act of resistance to colonial labor demands, but in a region that was reeling from recruitment, it undoubtedly won willing adherents. The administration repeatedly downplayed the uprising, but the Kongo-Wara was a clear indication of the fragility of French rule well into the 1930s.

The region that raised the most concern covered an area roughly the size of California—about 170,000 square miles—primarily in Ubangi-Shari but spilling into northern Middle Congo and Cameroon. Reports described a countryside racked with fear, anxiety, and uncertainty from recruitment-related violence. An administrator touring northern Middle Congo in late 1927 reported that following recruitment in one town, the neighboring villages cleared out: "Since Makoua, the villages we passed through (Dounga, Makounbé, Assagnan, Laboi) are deserts." At best, the official encountered a few chiefs and some women. Only one village chief presented workers to the recruiters; armed guards gathered the rest of the volunteers "by employing force." The violence did not end there, however, as rebellious villagers killed chiefs and soldiers involved in the recruitment.

In 1929 the inspector Jean Marchessou concluded that the violence in the region resulted from the fact that French rule had "never been established" there. He visited the area just as a process of "pacification"—that is, a forceful military campaign—had quelled some of the more recalcitrant villages. But the reality of France's weakness in the area was bleak. "Our occupation," he wrote, was "strained to the extreme," rendering it essentially "illusory." Four years after the Pacha Affair, vast subdivisions remained "often unoccupied" by Frenchmen. If an organized, regulated process was the goal, then the colony had little business continuing recruitment for the Congo-Océan considering its complete lack of authority. Calling for the establishment of "equilibrium" in the region, Marchessou recommended ending recruitment until villages could recover from the recent disturbances. The region had been left to its own devices for far too long; the French badly needed a "native policy in the true sense of the term."

It was in these circumstances—where French control ranged from "lamentable" to "illusory"—that low-level administrators were expected to find volunteers to work on a notoriously deadly railroad hundreds of kilometers away. The experiences of men like Georges Pacha illustrate how demands and pressure from Brazzaville set in motion events that often ended in violence. Many officers like Pacha, far from being brutalized, were young and inexperienced, working in parts of northern Middle Congo, Ubangi-Shari, and Chad where the French administration had only the most tenuous presence. When ordered to recruit hundreds of workers, Pacha's initial response had been disbelief. Like other officials, he pushed back, saying that finding so many workers risked destabilizing an already-troubled region. In response to this internal resistance, ranking officials sometimes lowered their expectations, or agreed that certain populations should be left out of recruitment. But the pressure remained clear: finding workers for the railroad was the priority.

The use of violence was not an inevitability: in areas where administrators had better relations with local chiefs and communities, or where they made more persuasive cases for the benefits of working on the railroad, men and women agreed to recruitment. Sadly, the prescribed method of recruitment did not always go according to plan. Where entire communities fled into the bush to avoid approaching recruiters, the search for "volunteers" devolved into a veritable manhunt. Where local chiefs refused to assist, administrators relied on armed African soldiers to intimidate, terrorize, and kill to meet quotas; they tied men at the neck to ensure their delivery to inspection camps; they razed villages to assert power. These were historic methods, tried and true since the golden era of the concessionary companies.

Considering the evidence that not all administrators in the field wanted to fulfill the quotas demanded by Brazzaville, it is worth asking how and why they were convinced to act, especially when brutality was a necessary part of the process. Administrators were certainly conditioned to follow orders. They were a part of a hierarchy that they had willfully joined and that they hoped to climb. To get out of their isolated regions of Equatorial Africa—widely considered among officers to be a backwater of the empire—officers needed to build a personnel file reflecting efficiency and diligence.

Again, racial esprit de corps was a powerful incentive: orders from Brazzaville reflected the decisions and aims of a body that was institutionally and racially defined. Albert Memmi, the Tunisian Jewish philosopher who brilliantly analyzed the psychology of modern colonialism, observed that colonizers bought into the system because it was what buttressed their power and privileges. Sympathy for Africans rarely trumped one's service to colonial power. "Wonder has been expressed," Memmi pointed out, "at the vehemence of colonizers against any among them who put colonization in jeopardy." To refuse to follow orders was to be insubordinate, but also to be a traitor—to the colony and to the racial bonds that gave it meaning.

Officers were also motivated by the ideological goals of colonialism, including the role of the Congo-Océan in fulfilling French plans to develop the region. In the face of such widespread conflict, many a rational leader would have seen the colony as ill prepared to build a railroad, but Antonetti saw it differently. The railroad promised to solve all the colony's troubles, from its economic backwardness to its political instability. Indeed, Antonetti even insisted it would—paradoxically—minimize colonial violence. Improved French rule required better transit and communication systems, he emphasized, "which depend themselves on the construction of the railroad." His argument was dizzying in its circularity: it essentially assured that the violence caused by the construction of the railroad would help alleviate the violence caused, overwhelmingly, by the construction of the railroad. Antonetti's faith in the curative possibilities of his railroad defied logical assessment.

Dizzying or not, Antonetti's defense of the centrality of the railroad to the colony's future motivated his men far and wide. Administrators were repeatedly reminded that the governor-general had christened the construction of the Congo-Océan *the* primary concern of the colony. When administrators resisted demands for workers, headquarters responded less with insistence than with explanation: "I first have to remind you that the construction of the railroad must come before all other interests." Studies by social psychologists on the ways officials function within bureaucratic hierarchies suggest that persuasion can be a far more effective means of motivating compliance than simply issuing orders. Echoing the colony's faith in the importance of recruitment not only encouraged men to join the common cause of colonization, it offered reassurance, professional as well as moral, that should recruitment turn violent, responsibility would fall on the colony, not on the individual administrator.

If a sense of common cause offered reassurance, so did the very isolation of many administrators. For decades, regional officials in Equatorial Africa had remarkable independence in ruling the countryside as they saw fit. The Indigénat, the code that defined colonial legal power, relied on administrators to police, judge, and punish a wide array of violations, from disorderly conduct to failing to report outbreaks of animal diseases. In his extensive report on the "the great suffering" of Equatorial Africa, René Maran argued that the broad and vague language of the Indigénat allowed for widespread abuse of power. As regional officials enjoyed total power in enforcing the law, it often resulted in punitive campaigns against communities that failed to pay taxes, follow orders, or show sufficient respect.

This spirit of independence informed official action in the age of recruitment. The lack of oversight, the slowness of correspondence, and—importantly—the vague, euphemistic language of administrative orders encouraged officers in the field to rely on their instincts without fear of being caught. The use of language like being "tough" with the indigenous population, and accepting whatever "regrettable consequences" might occur, meant that no means of coercion needed to be discounted. It was a tenet of administration in northern Middle Congo and Ubangi-Shari that physical force was an essential tool. Antonetti concurred, saying that being too easy on indigenous communities only emboldened the "native mentality" with its "taste for battle and pillaging" and even cannibalism. Demonstrations of force, then, not only helped provide "volunteers" for the railroad; they taught longer-lasting lessons. The interests of empire were doubly served.

Should the brutality of local officials be exposed and condemned, as André Gide exposed that of Pacha, liability could always
be displaced. The targets were usually Africans. The use of African soldiers allowed white officials, from the lowest administrator
to the governor-general, to blame atrocities on the alleged inherent
savagery of their men, mitigating—at least in their own souls and
minds—their moral and criminal culpability. The African guards
used to lead recruits to the south were understaffed, unsupervised,
and often without options for housing or feeding the recruits, yet
they assumed responsibility for the fates of their prisoners. In the case
of scandal, when a guilty party had to be sacrificed to the gods of
public opinion, African auxiliaries were a favorite option. Time and
again white administrators were cleared and forgiven. The empire
was a place where white civil servants could, and in many cases did,
get away with murder.

As GROUPS OF RECRUITS moved southward, the pressure to build
the railroad in a timely manner continued to drive the white men
involved. The violence of the project became less about physical
abuse than about deprivation. Assault and mistreatment certainly
didn't disappear. But what is equally important to consider is why
even men who were generally moral and emotionally balanced participated in a project that created such widespread misery. The dehumanization of the workers, which started on the long march to the
line, made brutality more acceptable and common and also obscured
workers' suffering more generally. Indeed, many Europeans showed
themselves to be entirely apathetic about the lives of workers, be
they exhausted, injured, sick, or dying.

In addition to the marching of recruits under the watchful eyes
of armed men, transportation relied on privately owned shipping
companies that penned agreements with the colonial state. Com-

panies were asked to move as many recruits as possible, for as little money as possible, leading to the dangerous overcrowding of men and women on ill-equipped boats. When ship captains were criticized for packing men and women into holds or crowding them onto decks without toilets, cooking facilities, or protection from the elements, they refused to take responsibility and even expressed resentment at having the human cargo in the first place.

Similar attitudes could be found on the construction site itself, especially in the Mayombe. There the administration's pressure on the Batignolles translated into a work ethic that encouraged white employees on the line to get as much done with as few workers as possible. Rouberol and his cadre knew that to pull off an engineering feat with a small number of poorly fed, poorly housed, and poorly managed men was impossible without deadly consequences. And yet no surplus of sympathy was shown for workers. Workers' experiences—especially their hardships—were remarkably absent in the vast bulk of documentation and correspondence that the Batignolles produced. When they did appear, it was often in the service of making a point about either the shortage of workers or their shortcomings. In 1926, for example, Albert Lebert, an executive at the Batignolles, reported to Antonetti that workers, because of a lack of food rations, went days without eating. Rather than expressing concern for the men's lives, Lebert's main motivation in writing was to establish that hunger led to illness, and illness to "a significant reduction in the performance of those who come to work."

If Batignolles executives showed a general lack of concern, the company's paperwork reflected similar values. The logbooks that the Batignolles kept of the construction effort were entirely bereft of detailed references to the health and welfare of workers. Preprinted forms distributed along the railway recorded weather conditions, which were dutifully noted with temperatures taken twice daily, as well as descriptions of the clouds in the sky. Similarly, the weekly tasks overseen by white employees, such as clearing forest, surveying the route, and leveling, were recorded down to the quantity of earth moved. By contrast, descriptions of the actual men doing the work were limited to two slim columns that registered the number of workers per day and total workdays completed for the week. A third small column demanded the same information for work completed by beasts of burden.

The company's response to a lacking workforce was to turn a blind eye to their agents' treatment of workers. The result was often a repertoire of yelling, slapping, and whipping that was entirely at odds with practices on construction sites in Europe; but such abuse was tacitly accepted and even defended by the Batignolles' leadership. One means of motivating men with intimidation without exposing white employees to criticism was the use of African *capitas* to drive men into action. By allowing *capitas* and guards to intimidate and punish workers, white employees kept a comfortable distance from the brutality; the resulting violence was, as many whites noted, just Africans abusing Africans. Some European men even shook their heads at such ugliness, convinced of the savage brutality of these alleged "primitives," feeling entirely absolved of responsibility.

And, of course, in cases where the violent party was a white overseer, he could rely on the support of the Batignolles, just as administrators in the north were assured that their superiors would defend the tactics used for recruiting men. When white employees were accused, the company regularly pointed to the overly zealous investigations of the Labor Service that interfered with the timely construction of the railroad. Company representatives openly and repeatedly bemoaned the prohibition of using physical force to get their men to work harder and faster. One more senior employee complained to Antonetti about having "no means of coercion" when dealing with workers, and blamed the Labor Service for undermining company agents by putting them "on the same footing" as the workers. Inevitably, the company argued that investigations of white employees only slowed progress on the railroad line.

Brutality itself was but one aspect of the indifference that reigned in the hearts and minds of many Europeans on the construction site. Albert Londres, who had experienced firsthand the horrors of the First World War, was particularly gifted at capturing the coldheartedness of the white employees he encountered. On his tour of the Mayombe, he met a *surveillant* who admitted to forcing his men to work even when they were sick. Like so many Europeans on the Congo-Océan, the *surveillant* pinned the blame on the workers. The problem, the man explained, was that Africans were weak. "You see," he said, "a black man wilts like a flower! In the evening he is in good health; the next day he trembles; the third day he is out of his mind; the fourth, he's finished." At roll call, if workers complained of illness, he felt their foreheads for fever, but "I don't know anything; I'm not a doctor." Unconcerned, he sent them to work "with a kick in the . . ."

The men Londres encountered gave voice to an insensitivity toward workers that not only made violence more likely but also enabled Europeans to stand by as thousands starved, grew sick, and died. Many white men were convinced that the only ones responsible for the deaths of Africans were the Africans themselves. Even doctors often blamed workers for their own demise. Vocally critical of the handling of workers, Dr. Boyé still cast blame on them for their allegedly filthy habits and their inability to resist diseases. Those afflicted with cachexia, he explained, "weaken, fall prey to nostalgia, and let themselves die." When a group of recruits from Chad, specially chosen for their strength and good health, still died at alarming rates, Boyé found a way to shift blame to them: "That shows the fragility of this race to being uprooted." The outspoken Dr. Lefrou concurred, blaming mortality on workers' "incorrigible uncleanliness" and their refusal to take care of themselves when sick.

Men like Boyé and Lefrou both expressed outrage at working
and living conditions and hotly criticized the colony's utter unpre-
paredness to manage thousands of workers. They spent much of
their time on the line trying to understand causes of death and making recommendations for improvement. And yet neither could shake
the belief that ultimately workers died because of what Boyé called
their "state of physical and moral inferiority." There was extraordinary irony in the argument, considering that European workers died
at relatively high rates as well. But logic had no place when race was
an issue. It was powerful evidence that imperial ideology was more
than rhetorical flourishes for impassioned speeches; the assumptions
of empire penetrated minds and informed sympathies.

———

COLONIAL VIOLENCE CAN BE imagined as a sign of European power.
To cite Max Weber again, states are defined by their monopoly
of the legitimate use of force. Technological and organizational
advances allowed Europeans to use a handful of armed soldiers to
subdue entire regions and remake societies. In Equatorial Africa,
French power was on display, it seems to follow, when regional
administrators sent soldiers to kidnap villagers and force young men
into recruitment. Similar demonstrations of power, in the form of
armed guards and *capitas*, of white men with whips and clubs and
dogs, drove men daily on the constructions site to move earth, dig
trenches, and lay track. Confident of their power, one might say,
Europeans in the French Congo got men without tools to build a
modern engineering wonder.

But a closer look reveals a very different story. The effort to
build the Congo-Océan demonstrated that violence and indifference stemmed not from strength but from weakness. In vast swaths
of territory, the thin presence and disputed status of the French gave
officials scant leverage to persuade able men to volunteer to work on

a distant construction project, especially when local rumors deemed recruitment to be a death sentence. Villages that could not be won over, bullied, or bribed had to be coerced. Under immense institutional pressure, but with insufficient resources, administrators vied for power using the very means that other local ethnic groups had employed to establish influence for at least a century: alliances, agreements, taxation, and terror. French officials were a strange tribe of white men in funny hats far from home, trying to win respect. They were not triumphant colonizers; in vast swaths of Equatorial Africa, their methods looked little different from those of the recalcitrant groups they derided as nasty and savage.

While French officials tried to blame high rates of illness and mortality on the racial inferiority of workers, the obvious root cause of the loss of life was administrative mismanagement. The French state failed to invest the thought, time, and manpower to provide the infrastructure, medical systems, and oversight that would have been necessary to build the railroad without massive loss of life. The construction of the Congo-Océan under any circumstances would have been an extraordinary undertaking with many potential dangers, including the substantial loss of life. But as inspectors pointed out, the administration could have taken steps to improve workers' chances of survival. More efficient transport, more reliable food and water rations, real housing, and sufficient medical treatment all could have improved conditions and lowered mortality.

The story of the Congo-Océan certainly demonstrates all too clearly the devastating potency of assumptions of racial superiority. But the faith that France could accomplish the impossible, that it could transform the Cinderella of the empire, was also an essential theme in the tragedy. The decisions to start building a railroad without a map, to start recruiting without outposts, and to clear forests without means of delivering essential supplies revealed the delusional self-confidence of the colonial state. Apologists chose to believe that such arrangements were mere details, insignificant in comparison to the greatness of the project and the beneficence of France. Such hubris proved costly in African lives, but it also—ironically—enabled the Congo-Océan to go largely unchallenged by the many voices deriding it as a humanitarian disaster. Antonetti's grandiose claims and cynical denials demonstrated how easily officials could deflect even the most damning criticisms and accusations of malpractice. The allure of grandeur and progress, coupled with a trust in the efficacy of bureaucratic regulation, were too much for even the most skeptical European politicians to resist.

The narratives of empire that were spun in defense of the railroad were worrisomely effective at burying prolonged human misery, the deaths of thousands of people, and the disappearance of communities. They even helped Europeans claim credit for building a railroad across an unforgiving forest while simultaneously erasing the sacrifices of the tens thousands of men and women who made it possible. For the French who oversaw the Congo-Océan, white triumph would always discount African trauma. The narratives and ideals that animated liberal empires have not been entirely dismantled in the wake of decolonization. The voices and experiences of the men and women who worked on projects like the Congo-Océan—men and women with names like Fotiguela, Malemale, and Garassena—offer one line of defense against future hubris. In their stories is the wisdom needed to see the potential duplicity of those who confidently inspire with promises of transformation, wealth, and humanity.

# ACKNOWLEDGMENTS

IN WRITING THIS BOOK, I've incurred many unpayable debts to colleagues, friends, and loved ones. I'm glad to have the opportunity to express my sincerest appreciation.

Like all books based largely on primary research, this one would have been impossible without the help of archivists and librarians near and far. During my stay in the Republic of Congo, Brice Owabira and Raoul Ngokaba not only provided essential guidance, they were exceptionally welcoming hosts as well. I would also like to thank the staff at the national archives in Brazzaville for their attentive assistance. In France, I benefited greatly from the archivists and staffs of the Archives nationales d'outre-mer (Aix-en-Provence), the Archives nationales du monde du travail (Roubaix), the Bibliothèque nationale de France (Paris), the Bibliothèque de documentation internationale contemporaine (Nanterre), the archives of the Ministère des affaires étrangères (La Courneuve), and the Académie des sciences d'outre-mer (Paris). In Geneva, Bernhardine Pejovic guided me expertly through the Archives of the League of Nations, as did Renée Berthon at the International Labor Organization. Much closer to home, Sarah Sussman, the fine curator of the French and Italian collections at Stanford, has been unstinting with her time and expertise. Without the dedication of these folks, this book would have been impossible.

Four people—good friends and brilliant scholars all—have profoundly informed my understanding of key subjects in this book. They deserve special attention and particular gratitude. Robert

Crews, Sean Hanretta, Aishwary Kumar, and Edith Sheffer have been ideal colleagues: creative, insightful, and true. All four shared with me their knowledge of history. All four offered their unique perspectives on how to write about violence and suffering. Perhaps most important, all four helped me process and diagnose the deceptive and often nefarious ways in which people and institutions can deny, distract, and manipulate; craft their own narratives; and justify even the most unforgivable behavior—all central themes of this book. I cannot thank them enough for their comradery, perceptiveness, and reliability over the years.

Naomi Andrews, Edward Berenson, Robert Crews, Brooke Durham, Karyn Panitch, Steven Press, Edith Sheffer, and Yan Slobodkin graciously read either large portions or the entirety of the manuscript and provided detailed criticisms that improved the book immeasurably. It is impossible to name all the people with whom I had meaningful conversations or from whom I welcomed advice, suggestions, or support. But a partial list of those to whom I am indebted includes Nimisha Barton, Annette Becker, Julian Bourg, Mark Braude, Gordon Chang, Christopher Church, Julia Clancy-Smith, Joshua Cole, Alice Conklin, Sarah Curtis, Dan Edelstein, Babacar Fall, Sarah Farmer, Didier Fassin, Stefan Hoffman, Alexander Keese, Sandrine Kott, Elizabeth Marcus, Samuel Moyn, Deborah Neill, Nigel Penn, Richard Roberts, Davide Rodogno, Jennifer Sessions, Greg Shaya, James Sheehan, Jean-François Sirinelli, Tyler Stovall, Marie-Pierre Ulloa, and Owen White. I am humbled by their generosity.

During a critical year in the life of the project, a number of wise fellows at the Stanford Humanities Center encouraged me to refocus what was then a sprawling project onto one of my many cases, the Congo-Océan Railroad. I am especially grateful for the direction offered at that time by Attiya Ahmad, Joseph Boone, Fred Donner, Erika Doss, Stephanie Malia Hom, and Molly Pucci. James

Lock, James Holland Jones, and Jacques Pepin patiently engaged my questions and provided much needed direction on the origins of AIDS. Laurie Steinberg helped calculate the nutritional value of food rations. A number of people helped with research assistance over the years. My thanks to Justin Daniels, Brooke Durham, Elizabeth Jacob, Laura Monkman, and Madina Thiam.

Presentations at universities and conferences informed my thinking about empire, humanitarianism, and violence in crucial ways. I am especially thankful for questions asked and suggestions made at talks given at Columbia University, Sciences Po-Paris, the Université de Genève, the Graduate Institute of Geneva, the University of Toronto, Stanford University, the University of Pennsylvania, the University of Chicago, and the University of California, San Diego, as well as multiple meetings of the Society for French Historical Studies and the Western Society of French History.

In the years in which this book took shape, I had the honor of working with both doctoral students in my department as well as fellows in the Andrew W. Mellon postdoctoral program I codirected for nearly a decade with Lanier Anderson. These people have been constant sources of wisdom and inspiration, they have challenged me intellectually and rewarded me with their creativity and acumen. Many of them engaged my project in ways more rewarding than I can say. In particular, I thank Boris Atanassov, Anne Austin, Jon Connolly, Kate Elswit, Brendan Fay, Elizabeth Jacob, Alan Mikhail, Laura Monkman, Sabauon Nasseri, Luca Scholz, Adena Spingarn, Derek Vanderpool, Adam Zientek, and Karen Zumhagen-Yekplé. Getting to know and learn from these folks has been the greatest professional privilege.

Sandy Dijkstra, as well as the outstanding team at her agency, showed her famous enthusiasm for and faith in this book when I needed it most. Finding a home for it at W. W. Norton has been more rewarding than I ever could have hoped. John Glusman is a gracious and judicious editor; working with him has been a distinct pleasure. My thanks also go to Helen Thomaides and the rest of the wonderful staff at Norton for ushering the manuscript to publication. Rebecca Karamehmedovic provided essential assistance with photograph permissions. Erik Steiner produced the maps. Much of the research and writing of this book took shape in an office affordably rented to me by Pete Mulvihill and Kevin Ryan at Green Apple Books in San Francisco. I cannot imagine better landlords.

The importance of friends and family when writing a book like this cannot be understated. Some fulfilled the double duty of giving practical assistance as well as cherished companionship. Frank Murphy joined me in the Republic of Congo to ride the Congo-Océan; he was an ideal travel partner, even if his love of photography almost got us arrested. Marie Murphy supplied her expert opinions, often at a moment's notice, about tricky translations. And, in addition to always bringing good cheer, Arcène Bilégué has shared his knowledge of all things central African, including its languages, cultures, flora, and fauna. Other friends listened, encouraged, smiled, and laughed. Dan Ahern, Sasha Baguskas, Martha Curbow, Lillie and Belle Sanchez, Makiko and Les Wisner have all been regular dinner companions who have brought much-needed levity and joy.

As has always been the case, it is my family—Karyn, Nathaniel, and Henry—who have remained unflappable in the years of research and writing. It was no doubt naïve to believe that I could work on a topic as often troubling as this one without it seeping into my daily life. I tried to leave my research at the office, so to speak, away from a home with growing kids. But I know that in some ways, they lived with the Congo-Océan as much as I—something they never asked for. And yet they never complained (too much!) and were always ready with words of love, reassurance, and laughter when I needed it most. For that and for all the things I could never put into words, I dedicate this book to them.

# LIST OF ABBREVIATIONS

AEF     l'Afrique équatoriale française; French Equatorial Africa

AILO    Archives of the International Labor Organization, Geneva

ALDH   Archives of the Ligue des droits de l'homme, Paris

ALN     Archives of the League of Nations, Geneva

ANMT   Archives nationales du monde du travail, Roubaix, France

ANOM  Archives nationales d'outre-mer, Aix-en-Provence, France

ARC     Archives de la République du Congo, Brazzaville

BIT      Bureau international de travail

CFCO   Chemin de fer Congo-Océan

GG      Governor-General

LDH    Ligue des droits de l'homme; League of the Rights of Man

SCB    Société de construction des Batignolles

# NOTES

## INTRODUCTION

3  **Boats from Europe:** ANMT: 89 AQ 945: Société de construction des Batignolles, "Lettres générales aux chantiers," dossier 2. This dossier contains letters from the construction site addressed to "M le Chef de DIVISION à Pointe-Noire" from 1928 to 1934.

3  **"nearly limitless":** "Sauvons l'Afrique équatoriale française," *L'Afrique française,* no. 5 (May 1924).

5  **A second photograph:** ANMT: 89 AQ 1000: Photographies. "Jeune femme et travaileur malades." The image was taken at Mavouadi, Kilometer 61, and was an annex to a report by Rouberol, June 25, 1925.

5  **"aestheticization of suffering":** Ariella Azoulay, *Civil Imagination: A Political Ontology of Photography,* trans. Louise Bethlehem (New York: Verso, 2012), 1.

5  **this collection of photos of workers:** The Batignolles images, now housed at ANMT, are still open to researchers only with special permission.

6  **On one page:** ANMT: 89 AQ 1000: Photographies, "À la Chasse entre Pointe-Noire et Côte Mateva," August 5, 1924.

6  **two white men and two white women out hunting:** On hunting as an imperial pursuit, see John M. MacKenzie, *The Empire of Nature: Hunting, Conservation, and British Imperialism* (Manchester, U.K.: Manchester University Press, 1997); and in the French context, Eric T. Jennings, *Imperial Heights: Dalat and the Making and Undoing of French Indochina* (Berkeley: University of California Press, 2011), esp. chap. 5.

6  **a group of workers in the hot sun:** ANMT: 89 AQ 1000: Photographies, "Les Travaux au Kil. 53," March 2, 1925.

7  **"frugal, consisting chiefly":** Gabrielle M. Vassal, *Life in French Congo* (London: T. Fisher Unwin, 1925), 72.

8  **"a frightful consumer of human lives":** André Gide, *Voyage au Congo* (Paris: Gallimard, 1927), 223

8  **according to investigations conducted in the 1930s:** Gilles Sautter, "Notes sur la construction du chemin de fer Congo-Océan (1921–1934)," *Cahiers d'études africaines* 7, no. 26 (1967): 269. Sautter himself estimates that at least sixteen thousand died.

8  **Unofficial estimates were far higher:** Mario Azevedo, "The Human Price of Development: The Brazzaville Railroad and the Sara of Chad," *African Studies Review* 24, no. 1 (1981): 12. Azevedo cites a number of sources, including Gaston Bergery, General Jean Hilaire, and debates in the National Assembly.

10  **"and that cost us only":** *Journal du Peuple,* April 28, 1929.

10  **Workers died at the hands:** Jacques Pepin, *The Origin of AIDS* (Cambridge, U.K.: Cambridge University Press, 2011), esp. chap. 3.

10  **building the Pyramids of Giza:** Deaths at Giza are based on the excavations of cemeteries at the site. See Zahi Hawass and Ashraf Senussi, *Old Kingdom Pottery from Giza* (Cairo: American University Press, 2008), 15–17. On the number of workers involved, see Mark Lehner and Zahi Hawass, *Giza and the Pyramids: The Definitive History* (London: Thames and Hudson, 2017).

10  **"French period" of building the Panama Canal:** See David McCullough, *The Path Between the Seas: The Creation of the Panama Canal, 1870–1914* (New York: Simon and Schuster, 1978).

10  **White Sea–Baltic Canal:** Anne Applebaum, *Gulag: A History* (New York: Anchor Books, 2004), chap. 4.

11  **Burma-Siam Railway:** Neil McPherson, "Death Railway Movements," http://www.mansell.com/pow_resources/camplists/death_rr/movements_1.html.

11  **"a concentration camp":** René Maran, "André Gide and l'Afrique noire," trans. Mercer Cook, *Phylon* 12, no. 2 (1951): 167.

12  **None has attempted to explore how or why:** The finest studies of the railroad are Gilles Sautter's long article, "Notes sur la construction du chemin de fer Congo-Océan (1921–1934)," *Cahiers d'études africaines* 7, no. 26 (1967): 219–99; and Joseph Gamandzori, "Chemin de fer: villes et travail au Congo (1921–1953)," Ph.D. diss., University of Lille, 1987. Blandine Sibille and Tuan Tran Minh, *Congo-Océan: de Brazzaville à Pointe-Noire, 1873–1934* (Paris: Éditions Frison-Roche, 2010). is a beautifully illustrated general account of the railroad by two nonhistorians. A fine collection of documents on the controversy surrounding the railway can be found in Ieme van der Poel, *Congo-Océan: un chemin de fer colonial controversé* (Paris: Harmattan, 2006), 2 vols. Rita Headrick, *Colonialism, Health and Illness in French Equatorial Africa, 1885–1935* (Atlanta: African Studies Association Press, 1994) offers an unparalleled analysis of health concerns surrounding the railroad. On the Batignolles, see Anne Burnel, *La Société de construction des Batignolles de 1914 à 1939: histoire d'un déclin* (Geneva: Librairie Droz, 1995).

12  **racist, callous, and violent:** That empire was inherently violent is accepted by most reputable historians. The origin of this argument can be traced to Frantz Fanon, *The Wretched of the Earth,* trans. Richard Philcox (New York: Grove Press, 2007); and Aimé Césaire, *Discourse on Colonialism,* trans. Joan Pinkham (New York: Monthly Review Press, 2001), who cites the Congo-Océan as an example of colonialism's brutality. The number of more recent histories of colonial atrocities is vast; they include, for example, Benjamin Claude Brower, *A Desert Named Peace: The Violence of France's Empire in the Algerian Sahara, 1844–1902* (New York: Columbia University Press, 2009); Adam Hochschild, *King Leopold's Ghost: A Story of Greed, Terror, and Heroism in Colonial Africa* (Boston: Houghton Mifflin, 1998); Isabel Hull, *Absolute Destruction: Military Culture and the Practices of War in Imperial Germany* (Ithaca, N.Y.: Cornell University Press, 2005); Mark Cocker, *Rivers of Blood, Rivers of Gold: Europe's Conflict with Tribal Peoples* (London: Jonathan Cape, 1998); Mike Davis, *Late Victorian Holocausts: El Niño Famines and the Making of the Third World* (New York: Verso, 2001); Caroline Elkins, *Imperial Reckoning: The Untold Story of Britain's Gulag in Kenya* (New York: Henry Holt, 2005); Marc Ferro, ed., *Le Livre noir du colonialisme: XVI<sup>e</sup>–XXI<sup>e</sup>: de l'extermination à la repentance* (Paris: R.

Laffont, 2003); Sven Lindqvist, *"Exterminate All the Brutes": One Man's Odyssey into the Heart of Darkness and the Origins of European Genocide* (New York: New Press, 1996); and Bertrand Taithe, *The Killer Trail: A Colonial Scandal in the Heart of Africa* (Oxford: Oxford University Press, 2009).

12   **"technical expert":** On technical experts, see Frederick Cooper, *Africa Since 1940: The Past of the Present* (Cambridge, U.K.: Cambridge University Press, 2002), 88. On the shift to a more humanitarian vision of empire after the First World War, see Albert Sarraut, *La Mise-en-valeur des colonies françaises* (Paris: Payot, 1923); and Yan Slobodkin, "Empire of Hunger: Famine and the French Colonial State, 1867–1945," Ph.D. diss., Stanford University, 2018. On humanitarianism after the war, see Bruno Cabanes, *The Great War and the Origins of Humanitarianism, 1918–1924* (Cambridge, U.K.: Cambridge University Press, 2014), 16.

12   **"lethargic sleep":** Antonetti quoted in "On va construire un chemin de fer Brazzaville-Océan," *L'Echo de Paris,* September 24, 1924.

13   **denounced by Americans and Europeans:** The most influential book on the subject of colonial atrocities is Adam Hochschild, *King Leopold's Ghost: A Story of Greed, Terror, and Heroism in Colonial Africa* (Boston: Houghton Mifflin, 1998). For other books that examine colonial violence in the Belgian Congo, see Kevin Grant, *A Civilised Savagery: Britain and the New Slaveries in Africa, 1884–1926* (New York: Routledge, 2005); John Tully, *The Devil's Milk: A Social History of Rubber* (New York: Monthly Review Press, 2011); and David Van Reybrouk, *Congo: The Epic History of a People,* trans. Sam Garrett (New York: Ecco, 2014). On early critics, see Roger Casement's 1903 diary and 1904 Congo Report in Séamas Ó. Síocháin and Michael O'Sullivan, eds., *Eyes of Another Race: Roger Casement's Congo Report and 1903 Diary* (Dublin: University College Dublin Press, 2003), and E. D. Morel, *Red Rubber: The Story of the Rubber Slave Trade Flourishing in the Congo in the Year of Grace 1906* (London: T. Fisher Unwin, 1906).

13   **a narrative of atrocity and reform:** For a critique of Hochschild and other scholarship on European atrocities, see David M. Gordon, "Precursors to Red Rubber: Violence in the Congo Free State, 1885–1895," *Past and Present* 236 (August 2017): 133–68; and Nancy Rose Hunt, *A Nervous State: Violence, Remedies, and Reverie in Colonial Congo* (Durham, N.C.: Duke University Press, 2016).

14   **Self-preservation is too often a greater priority:** While this book does not engage directly with theoretical work on institutional organization, works useful to historians include Roger Friedland and Robert Alford, "Bringing Society Back In: Symbols, Practices, and Institutional Contradictions," in *The New Institutionalism in Organizational Analysis,* ed. Walter Powell and Paul DiMaggio (Chicago: University of Chicago Press, 1991), 232–63; and Charlotte Cloutier and Ann Langley, "The Logic of Institutional Logics: Insights from French Pragmatist Sociology," *Journal of Management Inquiry* 22, no. 4 (2013): 360–80.

16   **"crime of crimes":** See, for example, Mark Mazower, "Violence and the State in the Twentieth Century," *American Historical Review* 107, no. 4 (2002): 1158–78. On the transcendence of genocide in both historical research and politics, see Dirk Moses, *The Problem of Genocide: Permanent Security and the Language of Transgression* (Cambridge, U.K.: Cambridge University Press, 2021).

16   **"voices" of the workers:** Scholars of slavery have considered the methodological

challenges associated with writing enslaved people back into American history. For a small sampling, see Saidiya Hartman, *Scenes of Subjection: Terror, Slavery, and Self-Making in Nineteenth-Century America* (New York: Oxford University Press, 1997); Wendy Anne Warren, "'The Cause of Her Grief': The Rape of a Slave in Early New England," *Journal of American History* 93, no. 4 (2007): 1031–49; Walter Johnson, *River of Dark Dreams: Slavery and Empire in the Cotton Kingdom* (Cambridge, Mass.: Belknap Press, 2013); and Tiya Miles, *Ties That Bind: The Story of an Afro-Cherokee*, 2nd ed. (Berkeley: University of California Press, 2015).

17    **Former workers left oral accounts:** Azevedo, "Human Price of Development," 1–19.

18    **Bodily trauma renders language imprecise:** See Elaine Scarry, *The Body in Pain: The Making and Unmaking of the World* (New York: Oxford University Press, 1985).

19    **"a realm of achievement rather":** Ned Blackhawk, *Violence over the Land: Indians and Empires in the Early American West* (Cambridge, Mass.: Harvard University Press, 2006), 1.

19    **Historians of modern Europe:** For two prominent examples, see Christopher Browning, *Ordinary Men: Reserve Battalion 101 and the Final Solution in Poland* (New York: HarperCollins, 1992); and Jan Gross, *Neighbors: The Destruction of the Jewish Community of Jedwabne, Poland* (Princeton: Princeton University Press, 2001). Both books inspired conferences, articles, newspaper articles, and books that weighed, responded to, and in some cases challenged their central arguments.

19    **deadly history of the Congo-Océan:** The railroad is conspicuously absent, for example, from the Internet's surprisingly large number of lists of the "deadliest construction projects" in history. These lists regularly include the building of the Aswan Dam, which caused approximately 500 fatalities in the 1960s, and that of the Karakoram Highway, which counted nearly 900 deaths between 1959 and 1978. They also include notoriously deadly American construction projects, such as the Los Angeles Aqueduct, the Hawks Nest Tunnel, and Hoover Dam, with fewer than 100 official deaths in each case. It is worth noting that colonial projects are rarely listed.

19    **Such willful forgetfulness:** On national memory, see Stanley Cohen, *States of Denial: Knowing About Atrocities and Suffering* (Malden, Mass.: Polity Press, 2001).

21    **"pornography of pain":** Karen Halttunen, "Humanitarianism and the Pornography of Pain in Anglo-American Culture," *American Historical Review* 100, no. 2 (1995): 303–34.

21    **Africans are too commonly portrayed as passive victims:** Some of these concerns are eloquently discussed in Nancy Rose Hunt, "STDs, Suffering, and Their Derivatives in Congo-Zaire: Notes toward an Historical Ethnography of Disease," in *Vivre et penser le sida en Afrique,* ed. Charles Becker et al. (Paris: Éditions Karthala, 1999), 111–31.

21    **They *oblige* history:** On the obligation to move past "the unthinkable" and to confront the past, see Georges Didi-Huberman, *Images in Spite of All: Four Photographs from Auschwitz,* trans. Shane B. Lillis (Chicago: University of Chicago Press, 2012), 3.

## CHAPTER 1: REMAKING THE CONGO

23    **It was a forbidding territory:** See Patrick Bratlinger, "Victorians and Africans: The Genealogy of the Myth of the Dark Continent," *Critical Inquiry* 12, no. 1 (1985):

166–203; and Josiah Blackmore and James Green, "European Images of the Kongolese in Books," in Alisa Lagamma, ed., *Kongo: Power and Majesty* (New York: Metropolitan Museum of Art, 2015), 119–30.

23 **"Going up that river":** Joseph Conrad, "Heart of Darkness" (1899), in *Heart of Darkness and Other Tales,* ed. Cedric Watts (Oxford: Oxford University Press, 2002), 136.

24 **Early European visitors:** See Rita Headrick, *Colonialism, Health and Illness in French Equatorial Africa, 1885–1935* (Atlanta: African Studies Association Press, 1994), 10–11; and Lagamma, *Kongo: Power and Majesty.*

24 **the region had established diplomatic ties:** See Mostafa Minawi, *The Ottoman Scramble for Africa: Empire and Diplomacy in the Sahara and Hijaz* (Stanford, Calif.: Stanford University Press, 2016).

26 **tearing apart kingdoms and communities alike:** Kairn A. Kleiman, *"The Pygmies Were Our Compass": Bantu and Batwa in the History of West Central Africa, Early Times to c. 1900 C.E.* (Portsmouth, N.H.: Heinemann, 2003), chap. 6.

27 **Pierre Savorgnan de Brazza:** On Brazza's life, see Maria Petringa, *Brazzà, A Life for Africa* (Bloomington: AuthorHouse, 2006); and Idanna Pucci, ed., *Brazza in Congo: A Life and Legacy* (New York: Umbrage Editions, 2009).

27 **from Gabon to the northern bank:** Steven Press, *Rogue Empires: Contracts and Conmen in Europe's Scramble for Africa* (Cambridge, Mass.: Harvard University Press, 2017), esp. chap. 3; and Victoria Thompson and Richard Adloff, *The Emerging States of Equatorial Africa* (Stanford, Calif.: Stanford University Press, 1960), 6–8. See also Catherine Coquery-Vidrovitch, *Brazza et la prise de possession du Congo, 1883–1885* (Paris: Mouton, 1969).

27 **Brazza's expeditions transformed him:** On Brazza as hero, see Berny Sèbe, *Heroic Imperialists in Africa: The Promotion of British and French Colonial Heroes, 1870–1939* (Manchester, U.K.: University of Manchester Press, 2013), esp. 149–52.

27 **Brazza's distance and "rogue" diplomacy:** On the full ramifications of Brazza's agreements with Iloo, see Press, *Rogue Empires,* 108–16.

27 **the French Congo:** On Makoko Iloo, see Emmanuel K. Akyeampong and Henry Louis Gates, Jr., eds., *Dictionary of African Biography* (Oxford: Oxford University Press, 2012), 4:153–54.

28 **the Congo Free State:** Maurice Rondet-Saint, *Dans notre empire noir* (Paris: Société d'éditions géographiques, maritimes et coloniales, 1929), 93.

28 **"an apostle of liberty, the new apostle of Africa":** Stanley quoted in Edward Berenson, *Heroes of Empire: Five Charismatic Men and the Conquest of Africa* (Berkeley: University of California Press, 2011), 50. Berenson provides rich accounts of the two men's personae.

28 **appealed to a wide spectrum of Frenchmen:** Ibid., chap. 2. On the evangelical possibilities of Brazza's work, see Charles Verne, *La France au Congo et Savorgnan de Brazza* (Paris: Librairie Fischbacher, 1887).

30 **a particularly French vision of civilization:** On the civilizing mission, see Alice Conklin, *A Mission to Civilize: The Republican Idea of Empire in France and West Africa, 1895–1930* (Stanford, Calif.: Stanford University Press, 1997).

30 **"All those who touch our flag are free":** Brazza quoted in Blandine Sibille and Tuan Tran Minh, *Congo-Océan: de Brazzaville à Pointe-Noire (1873–1934)* (Paris: Éditions Frison-Roche, 2010), 27.

30 **He often used the word *humanité*:** See, for example, Napoléon Ney, ed., *Conférences*

*et lettres de P. Savorgnan de Brazza sur les trois explorations dans l'ouest africain de 1875 à 1886* (Paris: Maurice Dreyfous, 1887), 63, 67.

30 **Allusions to humanity in colonization:** See Naomi J. Andrews, "'The Universal Alliance of All Peoples': Romantic Socialists, the Human Family, and the Defense of Empire during the July Monarchy, 1930–1848," *French Historical Studies* 34, no. 3(2011): 473–502; and Andrews, "Breaking the Ties: French Romantic Socialism and the Critique of Liberal Slave Emancipation," *Journal of Modern History* 85, no. 3 (2013): 489–527. For broader accounts of *mise-en-valeur* and its ramifications for French republicanism, see Raymond Betts, *Assimilation and Association in French Colonial Theory* (New York: Columbia University Press, 1960); Conklin, *Mission to Civilize*; and Gary Wilder, *The French Colonial Nation-State: Negritude and Colonial Humanism between the Two World Wars* (Chicago: University of Chicago Press, 2005).

30 **"These cannibal tribes":** *Exposé présenté par M. P. Savorgnan de Brazza, dans la séance générale extraordinaire, tenue au Cirque d'hiver le 21 janvier 1886* (Paris: Société de géographie, 1886), 22.

31 **"developed thanks to [Brazza's]":** Verne, *France au Congo*, 6. It was a combination with remarkable staying power. In 2006, Brazza's remains were reinterred in a newly built mausoleum in Brazzaville that celebrates the memory of the explorer and conqueror for his "humanitarian and peace-loving ideas." (This quote is from the June 2015 "Exposition-photos: La vie Pierre Savorgnan de Brazza" at the Brazza mausoleum in Brazzaville, Republic of Congo.)

31 **"Without a railroad":** Countless sources on the origins of the Congo-Océan mention Stanley's phrase, though none offers a specific reference. There is agreement, however, that Stanley made the observation before Brazza despite the fact that a Belgian train, from Léopoldville to Matadi, would not be started until 1890.

32 **"enormous riches":** Gilles Sautter, "Notes sur la construction du chemin de fer Congo-Océan (1921–1934)," *Cahiers d'études africaines* 7, no. 26 (1967): 220n1.

32 **"the vastest ensemble of navigable rivers":** Savorgnan de Brazza quoted in "Le Chemin de fer 'Congo-Océan,'" *Presse coloniale* 18 (October 1925): 7.

32 **"routes of penetration":** The assessment is Catherine Coquery-Vidrovitch's: "Les Idées économiques de Brazza et les prémières tentatives de companies de colonisation au Congo Français, 1885–1898," *Cahiers d'études africaines* 5, no. 17 (1965): 58.

32 **Brazza's vision to remake Equatorial Africa:** Ibid., 59.

32 **Expansion to the northwest:** Thompson and Adloff, *Emerging States of Equatorial Africa*, 10.

33 **a handful of stations and posts:** Phyllis M. Martin, "The Violence of Empire," in *History of Central Africa*, ed. David Birmingham and Phyllis M. Martin (New York: Longman, 1983), 2:10.

33 **The reigning ideology of the era:** Paul Leroy-Beaulieu's *De la colonisation chex les peuples modernes* (Paris: Félix Alcan, 1908) went through six editions between its publication in 1874 and the early twentieth century.

33 **difficulty finding serious investment:** See Catherine Coquery-Vidrovitch, *Le Congo au temps des grands compagnies concessionnaires, 1898–1930* (Paris: Mouton, 1972), 30–46.

33 **Brazza had suggested a combined:** Sautter, "Notes sur la construction," 220–21.

33 **the perils of early exploration:** "Le Capitaine Pleigneur mort sur le Niari," *Revue géographique international* 13, no. 148 (1888): 39.

**34** **"all the twists and turns of the route"**: Léon Jacob, "La Vallée du Niari-Kouilou," *Bulletin de la Société de Géographie* 15, 7th ser. (1894): 407–8.

**34** **A railroad would then be built**: Sautter, "Notes sur la construction," 221–22.

**34** **the government was already investing**: Henri Auric, *L'Avenir du Congo et le "Congo-Océan"* (Paris: Presses modernes, 1928), 23–24. On other missions, see Sautter, "Notes sur la construction," 223–24.

**35** **"our interests are relatively weak"**: Quoted in Coquery-Vidrovitch, *Congo au temps des grands compagnies*, 31.

**35** **His tenure had failed to build infrastructure**: Headrick, *Colonialism, Health and Illness*, 18.

**35** **rubber boom led to unknown levels of speculation**: Coquery-Vidrovitch, *Congo au temps des grands compagnies*, 46.

**35** **international agreements established**: Whether the concessionary system was even in keeping with the Berlin Congress caused diplomatic tension with Britain. See S. J. S. Cookey, "The Concession Policy in the French Congo and the British Reaction, 1898–1906," *Journal of African History* 7, no. 2 (1966): 263–78.

**36** **In the Congo Free State, concessions**: Coquery-Vidrovitch, *Congo au temps des grands compagnies*, 49.

**37** **The white men—and they**: Ibid., 78.

**37** **most of the concessions faced financial ruin**: Cookey, "Concession Policy in French Congo," 266.

**38** **"a taste for work"**: Quoted in Coquery-Vidrovitch, *Congo au temps des grands compagnies*, 118.

**38** **villages were being burned and fields razed**: Ibid., 126–27.

**39** **"an entertainment as new"**: "Les Bourreaux des noirs," *Le Matin* 22, no. 7662 (February 16, 1905): 1. For a more detailed account of the scandal, see Berenson, *Heroes of Empire*, chap. 6.

**39** **"A detonation rang out"**: Ibid.

**39** **Voulet and Chanoine were ultimately killed**: For more on the scandal, see Bertrand Taithe, *The Killer Trail: A Colonial Scandal in the Heart of Africa* (Oxford: Oxford University Press, 2009).

**40** **white men participating in various tortures**: *Assiette au beurre*, no. 206 (March 11, 1905).

**42** **his dispatches appearing in *Le Temps***: Catherine Coquery-Vidrovitch, ed., *Mission Pierre Savorgnan de Brazza. Commission Lanessan* (Neuvy-en-Champagne: Passage clandestin, 2016), 12–13.

**42** **The choice of Brazza**: See Berenson, *Heroes of Empire*, 198–200.

**42** **The stock-in-trade of popular journals**: See H. Hazel Hahn, "Heroism, Exoticism, and Violence: Representing the Self, 'the Other,' and Rival Empires in the French and English Illustrated Press, 1880–1905," *Historical Reflections/Réflexions Historiques* 38, no. 3 (2012): 62–83.

**43** **"The great explorer"**: *Le Petit Journal* (*supplément illustré*) 16, no. 748 (March 19, 1905): 89.

**43** **"without having fired a single shot"**: Ibid., 90.

**43** **Brazza's mission heard but**: Coquery-Vidrovitch, *Mission de Brazza*, 20–21; and Félicien Challaye in *Le Temps*, June 10, 1905. On Casement, see Seamas O'Siochain and Michael O'Sullivan, eds., *The Eyes of Another Race: Roger Casement's Congo Report* and *1903 Diary* (Dublin: University College Dublin, 2003); and *Africa. No. 1 (1904). Correspondence and Report from His Majesty's Consul at Boma Respecting the Administration of the Independent State of the Congo. Presented to both Houses of Parliament by Command of His Majesty* (London: Harrison and Sons, 1904).

43 **"veritable acts of cruelty":** Félicien Challaye, in *Le Temps,* September 26, 1905.

43 **ordered one of his African soldiers to summarily:** "Le Rapport Brazza," in Coquery-Vidrovitch, *Mission de Brazza,* 65.

44 **lacked sufficient evidence to reach a verdict:** Ibid., 75–81.

44 **"produce some rubber":** Coquery-Vidrovitch, *Congo au temps des grands compagnies,* 174–75.

44 **"les femmes de Bangui":** Ibid.

44 **Brazza found that the kidnapping:** Coquery-Vidrovitch, *Mission de Brazza,* 27–29.

45 **"veritable massacres of blacks":** ANOM: 2 AFFPOL 19: "Violences commises contre les indigènes au Congo," c. 1909.

45 **"When he had the energy to speak":** Félicien Challaye, "Le Congo français," *Cahier de la quinzaine* 7, no. 12 (1906): 17.

46 **"it is superficial, incomplete":** "Le Rapport Brazza," in Coquery-Vidrovitch, *Mission de Brazza,* 220.

46 **"apostles of development":** Émile Gentil in the introduction to Fernand Rouget, *L'Expansion coloniale au Congo français* (Paris: Emile Larose, 1906), iv.

46 **In its place, Lanessan produced:** Coquery-Vidrovitch, *Mission de Brazza,* 19–22.

47 **It was not "rediscovered" until 1965:** Coquery-Vidrovitch found the Brazza report while researching her dissertation. On its nondisappearance, see her introduction, ibid.

47 **from equatorial rain forests to sun-scorched deserts:** On the administrative evolution of the colony, see Hopiel Ebiatsu, *Administrateurs, marchands et indigènes: le Congo sous domination colonial française au début du XXᵉ siècle* (Saint-Denis: Edilivre, 2013), 45–46.

47 **The number of administrators:** Coquery-Vidrovitch, *Congo au temps des grands compagnies,* 85.

48 **a French company made Mindouli:** Auric, *L'Avenir du Congo,* 25–26.

48 **Bel nonetheless insisted that the train:** Pamphlet, "Mission Bel au Congo français. Séance du 7 Février 1908," *La Géographie,* Bulletin de la Société de géographie (Paris: Masson et Cie., 1908).

49 **Société de construction des Batignolles:** Anne Burnel, *La Société de construction des Batignolles de 1914 à 1939: histoire d'un déclin* (Geneva: Librairie Droz, 1995), 33.

49 **The price tag was estimated at 95 million francs:** Auric, *L'Avenir du Congo,* 27–31.

49 **"a wave of skepticism and also of discouragement":** *Les Annales coloniales,* February 28, 1922.

50 **Colonial reformers thought in terms:** On the notion that the rest of the world "lacked" what modern Europe possessed, see Dipesh Chakrabarty, "Postcoloniality and the Artifice of History: Who Speaks for 'Indian' Pasts?" *Representations* 37 (Winter 1992): 1–26.

50 **saw infrastructure as key:** On development in other empires, see Suzanne Moon, *Technology and Ethical Idealism: A History of Development in the Netherland East Indies* (Leiden: CNWS Publications, 2007); and Joseph M. Hodge, *Triumph of the Expert:*

*Agrarian Doctrines of Development and the Legacies of British Colonialism* (Athens: Ohio University Press, 2007).

**50**  **similar projects would pull colonial populations:** Conklin, *Mission to Civilize*;, and Eugen Weber, *Peasants into Frenchmen: The Modernization of Rural France, 1870–1914* (Stanford, Calif.: Stanford University Press, 1976).

**50**  *mise-en-valeur,* **or development:** Albert Sarraut's vision was outlined most forcefully in *La Mise-en-valeur des colonies françaises* (Paris: Payot, 1923). His chapter 6 deals extensively with infrastructure.

**50**  **"harmonious accord of races":** ANOM: GG AEF: 2H 6: Albert Sarraut, speech, November 5, 1923, 4.

**51**  **"developing our magnificent colonial domain":** Poincaré quoted in Burnel, *Société de construction des Batignolles,* 66.

**51**  **"human solidarity":** ANOM: GG AEF: 2H 6: Sarraut, speech, November 5, 1923, 13.

**51**  **Advocates of empire in the 1920s:** Grant, *Civilised Savagery,* 12.

**51**  **"In the name of humanity's right to life":** ANOM: GG AEF: 2H 6: Sarraut, speech, November 5, 1923, 10.

**51**  **a humanitarian narrative for empire building:** Thomas W. Laqueur, "Bodies, Details, and the Humanitarian Narrative," in *The New Cultural History,* ed. Lynn Hunt (Berkeley: University of California Press, 1989), 176–204.

## CHAPTER 2: THE RIGHT MAN FOR THE JOB

**52**  **"question of life or death":** Albert Sarraut, *La Mise-en-valeur des colonies françaises* (Paris: Payot, 1923), 419.

**52**  **efforts to end hunger, child mortality, and disease:** *Presse coloniale illustrée* 15 (July 1922).

**53**  **Augagneur's wife had the honor of breaking ground:** *Les Annales coloniales,* February 28, 1922; Blandine Sibille and Tuan Tran Minh, *Congo-Océan: de Brazzaville à Pointe-Noire, 1873–1934* (Paris: Éditions Frison-Roche, 2010), 63.

**53**  **"We could seek the best":** *L'Avenir du Congo Belge,* February 13, 1921, quoted in Henri Auric, *L'Avenir du Congo et le "Congo-Océan"* (Paris: Presses modernes, 1928), 36.

**53**  **"absolute shortage":** ANOM: FM AFFPOL 2856: Augagneur, "Note pour la direction de l'Afrique," August 7, 1920. These quotes come from 1920, but Augagneur had been organizing a recruitment effort led by a Captain Colomb since 1919.

**54**  **officials in Dakar were less than enthusiastic:** For the full correspondence of the mission, see ANOM: FM AFFPOL 2856: File "Recrutement en AOF (Capitaine Colomb)," c. 1920.

**54**  **between 2,000 and 8,000 workers would be needed:** On various agreements made between the company and the state, see Anne Burnel, *La Société de construction des Batignolles de 1914 à 1939: histoire d'un déclin* (Geneva: Librairie Droz, 1995), 178–89.

**55**  **"an ineffective and even barbarous":** Goffin, "Le Chemin de fer du Congo (1907)," quoted in ANOM: GG AEF: 3H 8: Pégourier to minister of colonies, October 15, 1926, emphasis in the original.

**55**  **"The rumors circulating":** ANOM: GG AEF: 3H 37: Charbonnier to head of Pool district, March 29, 1924.

55  **The rumors were based in fact:** ANOM: GG AEF: 3H 37: Marchand to Antonetti, April 17, 1924.

55  **force men to "volunteer":** ARC: GG AEF: 258: R. de Guise to lieutenant governor of Middle Congo, November 16, 1923.

55  **"by persuasion and not by coercion":** Augagneur quoted in ANOM: GG AEF: 3H 8: Pégourier to minister of colonies, October 15, 1926, 6.

55  **the security and livelihood of all laborers:** Ibid.

55  **"a state of lamentable physical decay":** ANOM: GG AEF: 3H 7: [Health Service inspector] Dr. Levet to head of Bureau of Political Affairs of General Government, May 1, 1924.

55  **"manifestly insufficient":** Ibid.

56  **"real fear" of being forced:** ANOM: GG AEF: 3H 7: Marchessou to governor-general, c. September 1926.

56  **more and more men were needed:** Gilles Sautter, "Notes sur la construction du chemin de fer Congo-Océan (1921–1934)," *Cahiers d'études africaines* 7, no. 26 (1967): 242–43.

56  **"a wretched country":** "Faut-il poursuivre la construction du chemin de fer de l'AEF?" *La Dépêche coloniale et maritime* (April 13, 1923). *La Dépêche coloniale* was particularly critical of the project. See the collection of clippings at ANMT: 89 AQ 995: "Articles de Journaux."

56  **"quartzose and granite crags":** "La Grande Pensée de M. Augagneur coûtera cher en AEF," *Le Courrier colonial* (April 18, 1924).

57  **"one of the world's great railroads":** ANOM: AFFPOL 2856: Antonetti quoted in René Thierry, *L'Afrique équatoriale française et le chemin de fer de Brazzaville à l'océan* (Paris: Comité de l'Afrique française, 1925), 7.

57  **"the Governor of the Congo-Océan":** "L'Afrique équatoriale française en plein essor," *La Revue des ambassades et des questions diplomatiques et coloniales* (November 1936): 23.

57  **He was born in Marseille:** The most complete biographical account is in *Hommes et destins* (Paris: Académie des sciences d'outre-mer, 1988), 8:3–5. The entry seems to have been based largely on his personnel record, ANOM: EE II: 3015 (1): Antonetti.

58  **"great administrator":** *L'Homme libre* (April 8, 1938): 3; and "Mort de l'ancien gouverneur général," *La Croix,* April 9, 1938, 1.

58  **The *Journal des Débats* wrote a bit more:** "Deuil," *Journal des débats* (April 9, 1938), 2.

58  **major reference guides:** See, e.g., Eugène Guernier, ed., *L'Afrique équatoriale française. Encyclopédie coloniale et maritime* (Paris: Encyclopédie de l'Union française, 1950), ix.

60  **he was as much a product:** On the relationship between individuals and imperial policies and practices, see Martin Thomas, ed., *The French Colonial Mind* (Lincoln: University of Nebraska Press, 2012), vols. 1 and 2.

60  **entering the colonial service at age nineteen:** "Deuil," *Journal des débats* (April 9, 1938), 2.

60  **He received no advanced academic or professional training:** On the multiple means of colonial professionalization, see Pierre Singaravélou, *Professer l'empire: les "sciences colonials" en France sous la IIIᵉ république* (Paris: Publications de la Sorbonne, 2011).

**60** **The possibility of advancement:** Cédric Chambru and Scott Viallet-Thévenin, "Mobilité sociale et empire: les gouverneurs coloniaux français entre 1830 et 1960," *Revue d'histoire moderne et contemporaine* 66, no. 4 (2019): 53–88.

**60** **"hellish land":** Georges Hardy, *Histoire sociale de la colonisation française* (Paris: Larose, 1957), 194–95.

**61** **He served short spells across West Africa:** He was in Dahomey on two occasions, for six months in 1909–10 and for three months in 1911); in Senegal for a year and a half starting in 1914; and in Upper Senegal and Niger for less than a year in 1916–17.

**61** **a post he retained until his retirement:** *Hommes et destins*, 3.

**61** **"depositories of excrement":** On the shortcomings of the colonial administration in Africa, see William B. Cohen, *Rulers of Empire: The French Colonial Service in Africa* (Stanford, Calif.: Hoover Institute Press, 1971). A. H. Canu ("excrement") and Georges Hardy ("crime") are quoted on p. 24. On the capabilities of secretariats-general, see pp. 27–29.

**62** **a remarkably diverse, if small and transitory population:** Hardy, *Histoire sociale de la colonisation*, 195.

**62** **little interest in seeking the opinions of subjects:** Raymond Leslie Buell, *The Native Problem in Africa* (New York: Macmillan, 1928), 983.

**62** **Men in Antonetti's position:** Cohen, *Rulers of Empire*, 61.

**63** **"by our porters, our copyists":** Antonetti quoted in Myron Echenberg, *Black Death, White Medicine: Bubonic Plague and the Politics of Public Health in Colonial Senegal, 1914–1945* (Portsmouth, N.H.: Heinemann, 2002), 64.

**63** **"get rid of him":** Ibid.

**63** **"grossly immoral":** G. Wesley Johnson, "The Ascendancy of Blaise Diagne and the Beginning of African Politics in Senegal," *Africa: Journal of the International African Institute* 36, no. 3 (1966): 251.

**63** **he argued *against* a policy:** Alice Conklin, *A Mission to Civilize: The Republican Idea of Empire in France and West Africa, 1895–1930* (Stanford, Calif.: Stanford University Press, 1997), 207.

**63** **"the sad alliance of misery":** Marc Michel, *Les Africains et la Grande Guerre: l'appel à l'Afrique (1914–1918)*, new ed. (Paris: Karthala, 2014), 53.

**63** **"fiercely individualist":** Antonetti quoted in Nicole Bernard-Duquenet, "Le Front populaire et le problème des prestations en AOF," *Cahiers d'études africaines* 16, nos. 61–62 (1976): 166.

**64** **All that really mattered was:** Antonetti quoted in Conklin, *Mission to Civilize*, 232.

**64** **"a race that poverty and privation":** Raphaël Antonetti, *La Côte d'Ivoire, porte du Soudan* (Paris: Comité de l'Afrique française, 1921), 3, 11.

**64** **"the populations have always lived":** Antonetti speech, in *Inauguration du chemin de fer Congo-Océan. Pose de la première pierre du Port de Pointe-Noire* (n.p., c. 1934), 17.

**64** **Civilization was France's gift to its colonial subjects:** For a sense of forced civilizing in West Africa, see Conklin, *Mission to Civilize*, esp. chap. 7.

**65** **attitudes that emerged in the wake of the First World War:** See Raymond Betts, *Assimilation and Association in French Colonial Theory* (New York: Columbia University Press, 1960); and Gary Wilder, *The French Colonial Nation-State: Negritude and Colonial Humanism Between the Two World Wars* (Chicago: University of Chicago Press, 2005).

**65** **"our solicitude for the black race":** Antonetti, *Côte d'Ivoire, porte du Soudan*, 15.

**65** **"grains, local textiles were bought":** Ibid., 4.

66 **"change the economic life of several million men"**: Ibid., 7.
66 **"mediocrely fertile but well populated"**: Ibid., 10–12.
66 **"very lazy and incapable of efforts"**: Ibid., 14–15.
67 **"the French duty of humanity"**: Antonetti quoted in "Afrique équatoriale française," *Le Temps* (July 17, 1924), in ANMT: 89 AQ 995.
67 **"Today a new era opens for the Congo"**: "Pour la mise en valeur de l'Afrique équatoriale," *La Dépêche coloniale* (September 24, 1924).
67 **"To open for free trade access"**: ANOM:AFFPOL 2856: M. R. Antonetti, *Discours prononcé à la séance d'ouverture du Conseil de Gouvernement, Session ordinaire de décembre 1925* (Brazzaville: Gouvernement général, 1926), 7.
68 **"a turning point in the economic history"**: Ibid., 11–14.
68 **"dorsal spine"**: Ibid., 16.
68 **"the most serious scourge"**: Ibid., 18.

## CHAPTER 3: THE PACHA PRELUDE

70 **"From the Congo River to the Sangha"**: Albert Londres, *Terre d'ébène* (1927; Paris: Collection Motifs, 2006), 242–43, 246–47.
71 **"entire villages were *punished*"**: Ibid., 247. Such a view was echoed in administrative reports, as well. See, for example, ANOM: GG AEF: 3H 8: Pégourier to minister of colonies, October 15, 1926, 6.
72 **When chiefs did not seem to be allies:** For an extensive discussion of the role of chiefs in recruitment, see Joseph Gamandzori, "Chemin de fer: villes et travail au Congo (1921–1953)," Ph.D. diss., University of Lille (1987), 170–80.
72 **"the construction of the railroad must"**: On Marchand's career, see ANOM: EE II: 1119: Marchand personnel file.
72 **five hundred men:** ANOM: GG AEF: 5D 60: [Labor Service director] Marchand to head of Lower Ubangi district, February 24, 1925.
73 **Two riverboats made the trip:** Ibid.
73 **"thought it necessary to respectfully"**: ANOM: GG AEF: 5D 60: Augias to governor of Middle Congo, June 25, 1925.
73 **"maladroit and dangerous"**: ANOM: GG AEF: 5D 60: Pacha to head of district (Dongou), September 21, 1925.
74 **"whatever may be the regrettable"**: ANOM: GG AEF: 5D 60: Augias to head of N'Gotto subdivision, October 8, 1925.
74 **he said he'd be happy with fifty:** ANOM: GG AEF: 5D 60: Marchand to administrator of Lower Ubangi district, November 14, 1925.
74 **Gide accepted an invitation:** Alan Sheridan, *André Gide: A Life in the Present* (Cambridge, Mass.: Harvard University Press, 1999), xv.
74 **Gide had for decades wanted to travel there:** Ibid., 401.
74 **he dedicated his 1927 account:** André Gide, *Voyage au Congo* (Paris: Gallimard, 1995).
75 **"felt very lonely there"**: Joseph Conrad, "Stanley Falls, Early September, 1890," in *Heart of Darkness*, ed. Robert Kimbrough, 3d ed. (New York: W.W. Norton, 1988), 187.
76 **"sensual" attraction for "the negro race"**: Sheridan, *André Gide*, xv.

76  **both would be deeply disappointed:** On Gide's approach to Africa, from a literary perspective, see Christopher Lane, "'Savage Ecstasy': Colonialism and the Death Drive," *Discourse* 19, no. 2 (1997): 110–33; and Amar Acheraïou, *Rethinking Postcolonialism: Colonialist Discourse in Modern Literatures and the Legacy of Classical Writers* (New York: Palgrave Macmillan, 2008), esp. chap. 9.

76  **on the backs of the porters:** Gide expressed deep ambivalence about the *tipoye*. He found it uncomfortable, but also disliked perspiring, an inevitable consequence of walking in Equatorial Africa. Brett A. Berliner, *Ambivalent Desire: The Exotic Black Other in Jazz Age France* (Amherst: University of Massachusetts, 2002), 179–80.

76  **Semba assumed that Gide:** ANOM: GG AEF: 5D 60: Marchessou, deposition of Semba Gotto, Boda, December 3, 1925.

76  **the chief awakened Gide at two:** ANOM: GG AEF: 5D 60: Gide to governor-general, November 4, 1925.

76  **The villagers, however, did not want to leave their crops:** Ibid.

76  **Women and children were burned alive:** Ibid.

77  **"men, women, and children":** Marc Allégret, *Carnets du Congo: voyage avec Gide* (Paris: Éditions Presses du CNRS, 1987), 87.

77  **"poorly understood or poorly executed":** ANOM: GG AEF: 5D 60: Gide to governor-general, November 4, 1925.

77  **"condemned to circle around":** Ibid.

78  **"a large number of human lives":** Ibid.

78  **"thrown in a cell":** Ibid.

78  **"This unfortunate affair":** Ibid.

79  **The disturbing details uncovered:** On the impact of the reception of *Voyage au Congo,* see Berliner, *Ambivalent Desire,* 180–88.

79  **local witnesses were vague:** The interviewees include Bogobolo, Boubacara, and Bobindé as witnesses. According to the interview of Soh, the Bofi referred to Bobindé as "Botobolo." He also mentions three others burned to death, a month earlier, in the village of Dana. Local residents had different names than the administration for villages and properties. ANOM: GG AEF: 5D 60: *Procès-verbal,* Pacha deposition of Soh, November 22, 1925.

79  **The children left inside were burned alive:** ANOM: GG AEF: 5D 60: *Procès-verbal,* Marchessou and Pacha deposition of Soko, c. late November 1925.

80  **"willfully burned alive in a hut":** ANOM: GG AEF: 5D 60: *Procès-verbal,* Pacha deposition of Fotiguela, November 22, 1925.

80  **"Did the sergeant try to stop him?":** ANOM: GG AEF: 5D 60: *Procès-verbal,* Pacha deposition of Botobolo, November 22, 1925.

80  **"Neither the guard nor the sergeant":** ANOM: GG AEF: 5D 60: *Procès-verbal,* Pacha deposition of Babodji, November 22, 1925.

81  **Marchessou defined the old colonial hand:** ANOM: EE II: 2230: Marchessou personnel file.

81  **"I put [Ngounou] on the ground":** ANOM: GG AEF: 5D 60: Jacoulet, Tribunal of Lower Ubangi district, January 25, 1926.

81  **the Indigénat:** Gregory Mann, "What Was the *Indigénat*? The 'Empire of Law' in French West Africa," *Journal of African History* 50, no. 3 (2009): 331–53.

82 **"The responsibility for this affair":** ANOM: GG AEF: 5D 60: Marchessou to head of Lower Ubangi district, December 21, 1925.

82 **lost one-third of its population:** Ibid.

82 **"exoduses":** ANOM: GG AEF: 5D 60: Marchessou to AEF governor-general, December 22, 1925.

82 **"young and insufficiently experienced":** ANOM: GG AEF: 5D 60: Jacoulet to Antonetti, February 27, 1926.

82 **"paradoxically vast":** ANOM: GG AEF: 5D 60: Pacha, written statement, April 13, 1926.

82 **decorated veteran:** ANOM: Personnel colonial contemporain: EE II: 3796 (26): Georges Pacha.

82 **His reasons for choosing a life:** The mother of his son, also named Georges Pacha, had to write to the colonial minister and hire a lawyer to track him down. ANOM: Personnel colonial contemporain: EE II: 5128: Georges Pacha: Germaine Sister (Alais) to minister of colonies, March 22, 1924.

83 **"in open anarchy":** ANOM: GG AEF: 5D 60: Augias to AEF governor-general, December 8, 1925.

83 **"refractory":** ANOM: GG AEF: 5D 60: Augias to governor-general, January 17, 1926.

83 **"always very tough with the natives":** ANOM: GG AEF: 5D 60: Pacha written statement, April 13, 1926.

83 **"firm but fair":** ANOM: Personnel colonial contemporain: EE II: 8 APOM: Georges Pacha: review signed by Augias, September 15, 1925.

83 **"if he had the courage for it":** ANOM: GG AEF: 5D 60: Augias to governor-general, January 17, 1926.

84 **a local leader named Bapélé:** Ibid.

84 **claims of cannibalism:** ANOM: GG AEF: 5D 60: Augias to governor-general, December 8, 1925.

84 **"justice was severely exercised":** ANOM: GG AEF: 5D 60: Augias to governor-general, January 17, 1926.

84 **"I am not a torturer, far from it":** ANOM: GG AEF: 5D 60: Augias to governor-general, December 8, 1925.

85 **The French penchant for administrators:** Raymond Leslie Buell, *The Native Problem in Africa* (New York: Macmillan, 1928), 1:983.

86 **"unceasing difficulties":** ANOM: GG AEF: 5D 60: Augias to governor-general, December 8, 1925.

87 **about the size of the U.S. state of New Jersey:** Ibid.

87 **"very painful, in generally unhealthy":** Ibid.

87 **"painful period":** ANOM: GG AEF: 5D 60: Pacha written statement, April 13, 1926.

87 **the sheer loneliness:** The distance from Boda to Impfondo is just over 300 kilometers as the crow flies. Augias estimated it as about 500 kilometers with the bends and turns of rivers and paths. ANOM: GG AEF: 5D 60: Augias to governor-general, December 8, 1925.

88 **Pacha did not take kindly to the criticisms of André Gide:** The omelet story was recounted in ANOM: Personnel colonial contemporain: EE II: 5128: Georges Pacha: "Notes sur le passage de M. André Gide," October 9, 1926.

88 **"an exceptional situation implying":** ANOM: GG AEF: 5D 60: Pacha written statement, April 13, 1926.

88 **Pacha faced charges in two hearings:** ANOM: GG AEF: 5D 60: Prosecutor to governor-general, January 5, 1927.

89 **a further one-year suspended sentence:** ANOM: GG AEF: 5D 60: Augias to governor-general, December 30, 1926.

89 **he was moved by Pacha's youth and inexperience:** ANOM: GG AEF: 5D 60: Prosecutor to governor-general, January 5, 1927.

89 **He blamed the entire episode:** Correspondence about Augias's alleged responsibility for the violence committed by Pacha's men can be found in ANOM: 8APOM: 4: "Affaire Pacha."

89 **Pacha's men received:** ANOM: GG AEF: 5D 60: Head of Lower Ubangi district Jacoulet to Antonetti, c. January 1926.

90 **chose a hundred of them:** ANOM: GG AEF: 5D 60: Marchessou to governor-general, December 22, 1925.

90 **"physical constraints":** ANOM: GG AEF: 5D 60: Pacha written statement, April 13, 1926.

91 **"NON-VOLUNTARY needless to say":** ANOM: GG AEF: 5D 60: Amouroux to Henri Guernut, July 11, 1926.

91 **"manifestly no connection":** ANOM: GG AEF: 5D 60: Antonetti to LDH president, February 20, 1927.

91 **Niamba's atrocities had been committed:** Antonetti was still insisting that there was no connection between Pacha and recruitment two months after Marchessou's December 22, 1925, report to the governor-general.

92 **"The administration is the mosquito":** Londres, *Terre d'ébène,* 57–59.

## CHAPTER 4: MANHUNT

93 **"capable":** ANOM: GG AEF: 3H 7: Lefrou, "Détachement No. 55": Malemale, March 12, 1928.

94 **"insufficiently muscular legs":** Ibid.

94 **5,000 would make it from Chad:** These are the estimates according to Gilles Sautter, "Notes sur la construction du chemin de fer Congo-Océan (1921–1934)," *Cahiers d'études africaines* 7, no. 26 (1967): 259.

94 **A photograph from 1926:** Stanford University Special Collections: Militaire Afrique Oubangui-Chari Train Congo-Océan, 1926 recruitment.

96 **"cheap merchandise":** ANOM: GG AEF: 3H 25: "Circulaire au sujet du recrutement de la Main-d'Oeuvre," July 13, 1926.

96 **to pay his tired parents' taxes:** Emmanuel Dongala, *Le Feu des origines* (Paris: Albin Michel, 1987), 98–99.

97 **"You must tell them the truth":** "Circulaire au sujet du recrutement de la Main-d'Oeuvre," July 13, 1926.

97 **recruitment for the railroad sent men running:** Joseph Gamandzori, "Chemin de fer: villes et travail au Congo (1921–1953)," Ph.D. diss, University of Lille (1987), 179–80.

97   **the population around the railroad:** ANOM: GG AEF: 3H 9: Health Service inspector [Lasnet], *Mission d'inspection,* March 1, 1928. On Lower Congo ("The prospect of the construction"), the report quotes an official named Cercus.

97   **"All the fit men are hidden":** ANOM: AFFPOL 2856: Administrator Montchamp to governor of Chad, July 1, 1928.

97   **Rumors of high mortality rates:** See, for example, ANOM: GG AEF: 3H 8: Pégourier to minister of colonies, October 15, 1926, 10.

97   **"natives to hide their heads in their arms":** Gaston Bergery, *Air Afrique: voie impériale* (Paris: Bernard Grasset, 1937), 151.

98   **working on the Congo-Océan would force them:** Gamandzori, "Chemin de fer," 182–83.

98   **officials warned that the presence of soldiers:** ARC: GG AEF: 258: R. de Guise to Middle Congo lieutenant governor, November 16, 1923.

98   **"the *miliciens* took them":** Marcel Homet, *Congo: terre de souffrances* (Paris: Fernand Aubier, 1934), 936.

98   **Soldiers also intimidated uncooperative chiefs:** Homet recounted the story of a chief who was whipped with the *chicotte* for refusing to produce workers. Marcel Homet, "La Vérité sur l'AEF," *Esprit: revue internationale* 2, no. 18 (1934): 917, 920.

98   **Others stood up to administrators who asked:** ANOM: GG AEF: 3H 7: "Note sur la mortalité des travailleurs des chantiers du chemin de fer pendant l'absence du gouverneur général du 25 janvier au 19 octobre 1927," n.d.

98   **an unnamed chief from Boko-Sangho:** ANOM: GG AEF: 3H 8: Marchand, "Observations du Lieutenant-Gouverneur P.I. du Moyen-Congo sur le rapport du mois de juin concentrant le fonctionnement de la main-d'oeuvre du C.O.," c. 1925.

99   **they offered political adversaries as recruits:** Gamandzori, "Chemin de fer," 175–79.

99   **paying men to go in their stead:** ANOM: GG AEF: 3H 8: Marchand, "Observations du Lieutenant-Gouverneur P.I. du Moyen-Congo sur le rapport du mois de juin concentrant le fonctionnement de la main-d'oeuvre du C.O.," c. 1925.

99   **one guard was injured with a knife:** ANOM: AFFPOL 2856: Administrator Montchamp to governor of Chad, July 1, 1928.

99   **"There was death":** ANOM: AFFPOL 2856: Montchamp to governor of Chad, November 29, 1928. See also Gamandzori, "Chemin de fer," 197.

100  **"open resistance":** ANOM: AFFPOL 2856: Montchamp to Governor of Chad, November 29, 1928. See also Gamandzori, "Chemin de fer," 197.

100  **"Since Makoua, the villages":** H——, "Tournée du 21 novembre au 21 décembre 1927." The report is included as an appendix to Robert Poulaine, *Étapes africaines: voyage autour du Congo* (Paris: Nouvelle revue critique, 1930), 229–31.

100  **when recruitment took place in his region:** ANOM: GG AEF: 5D 47: "Lettre de Berbérati," June 9, 1928.

100  **reports of an exodus:** Gamandzori, "Chemin de fer," 197.

100  **"old colonist":** *Le Populaire,* August 9, 1934.

101  **"the brutalities, the exactions":** "Que se passe-t-il en Oubangui-Chari," *Les Annales coloniales,* January 10, 1925.

101  **"intensely scared of being sent":** ANOM: GG AEF: 3H 37: President of the

indigenous tribunal at Fort-Rousset, *Rapport sur l'Affaire Kondo, homicide volontaire. Jugement du 25 mai 1929, No. 10,* May 29, 1929.

102 **"in tracked animals about":** Pierre Contet, "Le Travail forcé au Congo français," *Le Populaire,* July 17, 1934, 1. The entire series of articles was collected by the colonial administration and is archived in ANOM: GG AEF: 5D 83.

103 **"I don't know":** Ibid., 2.

104 **three-quarters of all recruits should have been:** ANOM: GG AEF: 3H 8: Health Service inspector Boyé, *Rapport médical annuel (octobre 1924–décembre 1925),* 132.

104 **327 recruits arrived from Bangui with only 47 men able to work:** ANOM: GG AEF: 3H 9: Health Service inspector [Lasnet], *Mission d'inspection,* March 1, 1928.

104 **a young man named Gazaio:** ANOM: GG AEF: 3H 7: "Détachement No. 55.": Gazaio.

105 **"No. Mle. 8846":** This example comes from ANOM: GG AEF: 3H 8: Pégourier to minister of colonies, October 15, 1926, 82.

105 **sufficient garments were not available:** ANOM: GG AEF: 3H 7: "Circulaire," Paris, July 22, 1924.

105 **"painfully astonished":** ANOM: GG AEF: 3H 37: Antonetti to head of Likouala-Mossaka district, July 31, 1929.

105 **recruitment no longer required force or intimidation:** Sautter takes these claims at face value, but not all evidence in the archives suggest that recruitment had become entirely voluntary by 1929. Sautter, "Notes sur la construction," 251.

106 **"You have undone your rope":** ANOM: GG AEF: 3H 37: *Justice repressive indigène, Affaire Kondo: rapport sur l'affaire.*

106 **"incomprehensible":** ANOM: GG AEF: 3H 37: Antonetti to head of Likouala-Mossaka district, July 31, 1929.

106 **despite policies to the contrary:** ARC: GG AEF: 258: R. de Guise to lieutenant governor of Middle Congo, November 16, 1923.

107 **"mediocre results":** ANOM: GG AEF: 3H 37: Antonetti to head of Likouala-Mossaka district, July 31, 1929; and *Justice repressive indigène, Affaire Kondo: rapport sur l'affaire.*

107 **"In this country, the barges":** Londres, *Terre d'ébène,* 243–44.

108 **"a humane and rational colonization":** ANOM: AFFPOL 2856: Maran to minister of colonies, January 26, 1928. Maran's claim that the boats were intended for nonhuman cargo was corroborated in Londres, *Terre d'ébène,* 243–44.

108 **"If these passengers are dressed workers":** ANOM: AFFPOL 2856: Maran to minister of colonies, January 26, 1928.

108 **"criminal amalgam of counter-truths":** Ibid., emphasis in the original.

109 **"they have to feel the European":** ANOM: FM AFFPOL 2856: General Thiry, "Impressions de voyage," August 19–September 27, 1926.

109 **"The protection issue":** ANOM: FM AFFPOL 2856: Haugou, *Extraits du p.v. de la Commission permanente prévue à l'art. 14 du traité . . . passé à Paris le 15 juillet 1927, pour l'execution d'un service public de transports sur le Congo, l'Oubangui et la Sangha,* November 30, 1927.

110 **could not safely carry more than seventy-five workers:** Ibid.

111 **"going to have as a first result":** Ibid.

111 **208 workers pressed on board:** ANOM: FM AFFPOL 2856: Administrator Bonhomme, director of AEF Supply Corps, *Procès-verbal de constat,* February 24, 1928.

111 **"deplorable conditions":** ANOM: AFFPOL 2856: Ministry of Colonies to General Government (telegram), February 24, 1928.

111 **"excessive":** ANOM: AFFPOL 2856: Antonetti to Ministry of Colonies (telegram), February 28, 1928.

111 **steamboat suited for a handful of passengers:** On the size of the *Colonel Klobb*, see M. Rondet-Saint, *Sur les routes du Cameroun et de l'A.E.F.* (Paris: Société d'éditions géographiques, maritimes et coloniales, 1933), 117–18.

111 **"men, women, and children would sleep":** Marcel Homet, *Congo: terre de souffrances* (Paris: Fernand Aubier, 1934), 226.

112 **"such language and lack of politesse":** ANOM: GG AEF: 3H 7: General Agent Landeich to Deschamps, captain of *Colonel Klobb,* June 25, 1929.

112 **Annual average daily high temperatures:** Based on averages from Bangui from 2000 to 2012, in "Average Weather in Bangui," WeatherSpark, https://weatherspark .com/averages/29063/Bangui-Central-African-Republic.

112 **separation did not stop contagion:** ANOM: FM AFFPOL 2856: Antonetti to minister of colonies, August 25, 1928; enclosing report from Colonial Administrator Montacq to Antonetti, June 18, 1928.

113 **about 1,200 calories:** The estimate is based on 750 grams (slightly more than three cups) of rice per day and a 200-gram portion of dried fish. For information on caloric intake at Buchenwald, estimated at 1750 calories daily, see Mike Davis, *Late Victorian Holocausts: El Niño Famines and the Making of the Third World* (New York: Verso, 2001), 39. See also Atina Grossmann, "Grams, Calories, and Food: Languages of Victimization, Entitlement, and Human Rights in Occupied Germany, 1945–1949," *Central European History* 44, no. 1 (2011): 118–48.

113 **Some recruits could not tolerate the transport:** Marcel Homet, "La Vérité sur l'AEF," *Esprit: revue internationale* 2, no. 18 (1934): 922.

## CHAPTER 5: "THE MAYOMBE DOESN'T WANT US"

114 **"only a big, very ill-equipped village":** Marcel Homet, *Congo: terre de souffrances* (Paris: Fernand Aubier, 1934), 215.

114 **Brazzaville was more like a collection:** On the capital at this time, see Phyllis Martin, *Leisure and Society in Colonial Brazzaville* (Cambridge, U.K.: Cambridge University Press, 1995)

115 **those who were healthy were sent:** See, for example, ANOM: GG AEF: 3H 41: "Note sur l'organisation du camp d'entrainement," c. late 1927.

115 **The Mayombe is a low, densely forested mountain range:** Spelling of the region's name fluctuated, including *Mayumbe, Mayoumbe, Mayomba, Majombe,* and *Mayumba,* though *Mayombe* was most common and remains the accepted spelling of the region today.

115 **ridges and precipitous flanks . . . cast much of the region into darkness:** For a description of the region, see Ján Borota, *Tropical Forests: Some African and Asian Case Studies of Composition and Structure* (Amsterdam: Elsevier Science, 1991), 67–73.

116 **"the useless establishment":** Marcel Sauvage, *Les Secrets de l'Afrique noire* (Paris: Denoël, 1937), 50.

116 **"dumps between seven and thirteen"**: Ibid., 47–48.

116 **annual rainfall in the Mayombe:** Borota, *Tropical Forests,* 74.

117 **"hell of humidity"**: Sauvage, *Secrets de l'Afrique noire,* 50. Borota concurs that relative humidity rarely falls below 75 percent, in *Tropical Forests,* 74.

117 **"The virgin forest"**: Sauvage, *Secrets de l'Afrique noire,* 52–53.

117 **"no more than a skeletal generator"**: Ibid., 51.

117 **"You lose the measure"**: Ibid., 49.

117 **"the forest without joy"**: Ibid., 47.

117 **"a deadly weight seems"**: Gabrielle Vassal, *Life in French Congo* (London: T. Fisher Unwin, 1925), 72.

117 **"the forest closed around us like a tunnel"**: Albert Londres, *Terre d'ébène* (1927; Paris: Collection Motifs, 2006), 255.

117 **"dark vault"**: Robert Poulaine, *Étapes africaines: voyage autour du Congo* (Paris: Nouvelle revue critique, 1930), 30–31.

119 **"Dantesque chaos"**: Ibid., 70.

119 **"Apocalypse"**: J. Coupigny, preface to Michel R. O. Manot, *L'Aventure de l'or et du Congo-Océan* (Paris: Librairie Secrétan, 1946), 9.

119 **"brings the railroad"**: ANOM: FM AFFPOL 2856: Labor Service director, *Rapport sur le fonctionnement du Service de la main d'oeuvre de chemin de fer Congo-Océan (mois de juillet 1925),* August 16, 1925, 5.

119 **"chaotic configuration of the ground"**: ANOM: GG AEF: 3H 8: Pégourier to minister of colonies, October 15, 1926, 10.

119 **"The land of the Mayombe doesn't want us"**: Labor Service director, *Rapport sur le fonctionnement (mois de juillet 1925),* 17.

119 **drew the final path right through the region:** Gilles Sautter, "Notes sur la construction du chemin de fer Congo-Océan (1921–1934)," *Cahiers d'études africaines* 7, no. 26 (1967): 233–34.

120 **about 107,000 days per kilometer of track:** ANOM: GG AEF: 3H 8: Pégourier to minister of colonies, October 15, 1926, 2.

120 **"an unhealthy, humid forest"**: Alexandre Piquemal in *Journal officiel de la République française. Débats parlementaires. Chambre des députés,* June 14, 1929, 2061.

121 **they would be denied the proper funeral rituals:** ANOM: FM AFFPOL 2856: Governor-general to heads of districts and subdivisions, circular, c. 1926.

122 **workers found themselves separated:** Labor Service director, *Rapport sur le fonctionnement (mois de juillet 1925),* 13.

122 **The imposition of French was an immediate:** See André-Patient Bokiba, *Ecriture et identité dans la literature africaine* (Paris: Harmattan, 1998).

122 **"semblance of a human relationship"**: Primo Levi, *The Drowned and the Saved,* trans. Raymond Rosenthal (New York: Vintage, 1989), 91.

122 **"thirty different dialects"**: ANOM: FM AFFPOL 2856: Governor-general to heads of districts and subdivisions, circular, c. 1926.

123 **only three of eleven Europeans:** ANOM: FM AFFPOL 2856: Labor Service director, *Rapport sur le fonctionnement du Service de la main d'oeuvre de chemin de fer Congo-Océan (mois de mai 1925).*

123 **"It's no longer the Congo-Océan"**: Londres, *Terre d'ébène,* 258.

123 **"Bastards! Pigs!"**: Ibid., 254, 258.

123 **"careful regulation":** ANOM: FM AFFPOL 2856: "Note Personnelle pour le Ministre au sujet de l'organisation de la Main d'Oeuvre du chemin de fer Congo-Océan," 4 June, 1928.

123 **"treating workers with more":** Ibid.

124 **"particularly encouraged":** ANOM: GG AEF: 3H 7: *Circulaire relative aux mesures de protection sanitaire à appliquer sur tous les chantiers publics et privés de travailleurs indigènes dans toutes les colonies,* July 22, 1924.

124 **less than two francs a day for men:** ANOM: FM AFFPOL 2856: *Directives annuelles accompagnant le plan de recrutement, pour l'année 1930.*

124 **Pay for workers was also often withheld:** ANOM: GG AEF: 3H 37: *Plaintes de mauvais traitements formulées par les manoeuvres de Monsieur Caraslanis, entrepreneur (CFCO),* April 25, 1924; and ANOM: GG AEF: 3H 34: Head of Volante subdivision of Congo-Océan to entrepreneurs, "Note Circulaire," May 21, 1931.

124 **Camps were guarded by** *miliciens* **or** *gendarmes*: ANOM: GG AEF: 3H 9: Health Service inspector [Lasnet], *Mission d'inspection,* February 10, 1928.

124 **workers were usually "pushed" by armed guards:** ANOM: GG AEF: 3H 9: Health Service inspector [Lasnet], *Mission d'inspection,* March 9, 1928.

125 **"well organized":** On housing along the railroad, see the series of reports from February 1928 in ANOM: GG AEF: 3H 9: Health Service inspector [Lasnet], *Mission d'inspection,* February 20, 1928.

125 **most of the camps of the Mayombe:** ANOM: GG AEF: 3H 9: Health Service inspector [Lasnet], *Mission d'inspection,* March 9, 1928.

125 **Progress along the line was certainly slow:** See, for example, ANOM: FM AFFPOL 2856: *Schéma de l'roganisation des chantiers,* c. 1927.

125 **"our feelings of humanity":** ANOM: GG AEF: 3H 41: Labor Service director to governor-general, September 14, 1927.

125 **Large barracks could be subdivided into rooms:** The camp in M'Vouti was one of the few with separate housing for couples and families. ANOM: GG AEF: 3H 9: Health Service inspector [Lasnet], *Mission d'inspection,* February 20, 1928.

126 **"extreme euphemism for architecture":** ANOM: GG AEF: 3H 7: Jouillon, *Extrait de rapport d'inspection du médecin principal,* report no. 1H, Mission Pégourier, c. 1927.

126 **buildings attracted ticks, jiggers, and parasites:** ANOM: FM AFFPOL 2856: Boyé, *Rapport médical annuel (octobre 1924–décembre 1925),* 138; ANOM: GG AEF: 3H 9: Health Service inspector [Lasnet], *Mission d'inspection,* February 22, 1928.

126 **"straw huts where water penetrates":** Poulaine, *Étapes africaines,* 70–71.

126 **Roofs made of foliage rotted:** Health Service inspector [Lasnet], *Mission d'inspection,* February 20, 1928.

126 **they fashioned torches from straw:** ANOM: GG AEF: 3H 9: Health Service inspector [Lasnet], *Mission d'inspection,* February 10, 1928.

126 **"smoky and numb":** ANOM: GG AEF: 3H 7: Jouillon, *Extrait de rapport d'inspection du médecin principal,* report no. 1H, Mission Pégourier, c. 1927.

126 **workers used parts of the walls:** ANOM: FM AFFPOL 2856: Middle Congo administrative Affairs inspector, report, February 4–25, 1926, 15.

127 **camps in a derelict, unhygienic state:** ANOM: GG AEF: 3H 9: Health Service inspector [Lasnet], *Mission d'inspection,* February 20, 1928.

127 **Africans would refuse to wear clothing:** ANOM: FM AFFPOL 2856: Health Service inspector Condé to governor-general, December 14, 1926, 11.

127 **where workers had uniforms:** ANOM: GG AEF: 3H 9: Health Service inspector [Lasnet], *Mission d'inspection,* March 10, 1928.

127 **officials discussed supplying men with lunchboxes:** ANOM: GG AEF: 3H 8: Middle Congo lieutenant governor, "Observations" c. 1926.

127 **men and women receive a can:** ANOM: GG AEF: 3H 9: Health Service inspector [Lasnet], *Mission d'inspection,* March 10, 1928.

128 **"I have seen them":** ANOM: GG AEF: 3H 9: Health Service inspector [Lasnet], *Mission d'inspection,* February 5, 1928.

128 **methods of collecting and distributing water:** ANOM: GG AEF: 3H 7: AEF governor-general to Health Service inspector Lasnet, August 24, 1929.

128 **the faucet drained much of the camp's limited water supply:** ANOM: GG AEF: 3H 9: Dr. Bouffard to governor-general, November 20, 1931.

128 **"nothing will be neglected":** The order was part of the *arrêté* of January 20, 1925. ANOM: GG AEF: 3H 9: Health Service inspector [Lasnet], *Mission d'inspection,* March 9, 1928.

128 **Antonetti encouraged much-sought-after Sara recruits:** Mario Azevedo, "The Human Price of Development: The Brazzaville Railroad and the Sara of Chad," *African Studies Review* 24, no. 1 (1981): 8.

130 **"bewildering ease of negresses":** : Michel R. O. Manot, *L'Aventure de l'or et du Congo-Océan* (Paris: Librairie Secretan, 1946), 52.

130 **On at least one occasion widows were linked:** ANOM: GG AEF: 3H 8: Pégourier to minister of colonies, October 15, 1926, "Conclusions," 5.

130 **The exchange of wages, women's refusal:** See, for example, ARC: GG AEF: 361: *Tribunal indigène du 2ème degré de la circonscription du Bas Congo,* January 19, 1932.

130 **"neither drum, nor merriment":** ANOM: GG AEF: 3H 9: Health Service inspector [Lasnet], *Mission d'inspection,* February 20, 1928.

131 **the French could open little shops:** Ibid.

131 **Alcohol, one possible form:** Azevedo, "Human Price of Development," 6.

132 **Sports and games were out of the question:** ANOM: GG AEF: 3H 9: Health Service inspector [Lasnet], *Mission d'inspection,* February 20, 1928.

132 **"no moral assistance, no encouragement":** ANOM: GG AEF: 3H 9: Health Service inspector [Lasnet], *Mission d'inspection,* March 9, 1928.

132 **"at the pleasure":** ANOM: GG AEF: 3H 37: Charbonnier to head of Pool district, March 29, 1924.

133 **a single *meter* of laid rail required:** This average was taken over the course of the track from Kilometer 95 to 105. Henri Auric, *L'Avenir du Congo et le "Congo-Océan"* (Paris: Presses modernes, 1928), 63.

133 **His hands are blistered and throbbing:** Emmanuel Dongala, *Le Feu des origins* (Paris: Albin Michel, 1987), chap. 3. On Dongala, see Ieme van der Poel, *Congo-Océan: un chemin-de-fer colonial controversé* (Paris: Harmattan, 2006), 2:183.

134 **Across one stretch of less than ten miles:** Manot, *L'Aventure de l'or,* 34.

134 **thirty-six major viaducts:** Sautter, "Notes sur la construction," 236.

134 **the Batignolles relied almost exclusively on human power:** Anne Burnel, *La Société de construction des Batignolles de 1914 à 1939: histoire d'un déclin* (Geneva: Librairie Droz, 1995), 185.

135 **"almost without tools":** ANMT: 89 AQ 873: Antonetti to Rouberol, chief engineer of the SCB division, August 10, 1926.

135 **"the hands of blacks":** Londres, *Terre d'ébène,* 258–59.

135 **"Here, it's only the black man!":** Ibid., 245–46.

135 **three hundred men had been given only fifteen axes:** Labor Service director, *Rapport sur le fonctionnement (mois de juillet 1925),* 6.

136 **"methods were primitive in the extreme":** Gabrielle Vassal, *Life in the French Congo* (London: T. Fisher Unwin, 1925), 70.

136 **"replaced automatically and without charge":** Emile Vincent, in *Journal officiel de la République française. Débats parlementaires. Chambre des députés,* February 1, 1930, 359.

136 **"thousands of blacks":** Georges Nouelle, in *Journal officiel de la République française. Débats parlementaires. Chambre des députés,* January 28, 1930, 196.

136 **pushed the company to adopt basic equipment:** See, for example, ANMT: 89 AQ 873: Antonetti to Rouberol, November 4, 1926.

137 **the Batignolles responded accordingly:** Sautter, "Notes sur la construction," 238.

137 **In 1924 a large portion of supplies arrived:** Joseph Gamandzori, "Chemin de fer: villes et travail au Congo (1921–1953)," Ph.D. diss., University of Lille (1987), 94.

137 **"It's horribly costly, inhumane":** ANOM: FM AFFPOL 2856: Antonetti to minister of colonies, January 23, 1926.

137 **"exhausting and repugnant":** ANMT: 89 AQ 873: Antonetti to Rouberol, November 4, 1926.

137 **the only means of transport in the region:** ANMT: 89 AQ 873: Rouberol to Antonetti, August 19, 1926.

138 **"colonization capable of":** ANOM: GG AEF: 3H 7: "Note sur le portage en AEF," n.d.

138 **"abandon his tribe":** ANOM: GG AEF: 497: LDH president to minister of colonies, c. November 7, 1927.

138 **"chaotic and wooded terrain":** ANOM: GG AEF: 3H 8: Pégourier to minister of colonies, October 15, 1926, 86.

138 **porters were forced to ascend and descend steep passages:** ANOM: GG AEF: 3H 9: Health Service inspector [Lasnet], *Mission d'inspection,* February 22, 1928.

139 **it could take an hour or more:** ANOM: GG AEF: 3H 8: Pégourier to minister of colonies, October 15, 1926, 85.

139 **carry more than 25 kilos:** ANOM: GG AEF: 3H 9: Health Service inspector [Lasnet], *Mission d'inspection,* February 20, 1928.

139 **95-kilo load (over 200 pounds):** ANOM: GG AEF: 3H 8: Pégourier to minister of colonies, October 15, 1926, 84.

139 **"crushing and unmanageable burdens":** Ibid., 85. See also Labor Service director, *Rapport sur le fonctionnement (mois de juillet 1925),* 3.

139 **"excessively painful and dangerous":** ANMT: 89 AQ 873: Antonetti to [Rouberel], November 4, 1926.

140 **Hungry men often binged on their food quickly:** ANOM: GG AEF: 3H 9: Health Service inspector [Lasnet], *Mission d'inspection,* February 20, 1928.

140 **"phagedenic sores":** ANOM: GG AEF: 3H 37: Le Comte to AEF governor-general, July 11, 1929.

140 **"Makes no difference to me":** ANOM: GG AEF: 3H 36: Acting Governor-General Joseph-François Reste to SCB director, September 8, 1927.

140 **"I don't give a d—":** Ibid.

140 **"It will not escape you ":** Ibid.

141 **"indiscretions":** ANOM: GG AEF: 3H 9: Health Service inspector [Lasnet], *Mission d'inspection,* February 20, 1928.

141 **"a *capita* slapped them":** Londres, *Terre d'ébène,* 254–55.

141 **The mortality rate jumped to 83 percent among porters:** ANOM: GG AEF: 3H 8: Pégourier to minister of colonies, October 15, 1926, 83. Pégourier quotes a report by Hardy de Perini.

142 **"that provides the largest contingent":** ANMT: 89 AQ 873: Governor-General [Antonetti] to head of SCB division [Rouberol], July 21, 1926.

142 **"All those with some knowledge":** Henry Fontanier, in *Journal officiel de la République française. Débats parlementaires. Chambre des députés,* November 23, 1927, 3177.

142 **"Above all, porterage":** Marius Moutet, in *Journal officiel de la République française. Débats parlementaires. Chambre des députés,* June 14, 1929, 2,056.

142 **Paying greater attention to the psychiatric needs:** On colonial psychiatry, see Richard Keller, *Colonial Madness: Psychiatry in French North Africa* (Chicago: University of Chicago Press, 2007).

143 **In 1925 a medical inspector estimated:** ANOM: GG AEF: 3H 8: Boyé, *Rapport médical annuel (octobre 1924–décembre 1925).*

143 **assigned to a makeshift morgue:** Azevedo, "Human Price of Development," 13.

143 **falls, of being crushed in cave-ins, and slides:** Rita Headrick, *Colonialism, Health, and Illness in French Equatorial Africa, 1885–1935* (Atlanta: African Studies Association Press, 1994), 287.

143 **"a stillborn railroad":** Raymond Susset, *La Vérité sur le Cameroun et l'Afrique équatoriale française* (Paris: Nouvelle revue critique, 1934), 131.

144 **"The result of this contempt":** Ibid., 135.

144 **"but it must be said":** Ibid., 136.

## CHAPTER 6: TROPIC OF CRUELTY

146 **"monopoly of the legitimate use":** Max Weber, "Politics as a Vocation," in *From Max Weber: Essays in Sociology,* trans. H. H. Gerth (Oxford: Oxford University Press, 1946), 78.

146 **"the willful inflicting of physical pain":** Judith N. Shklar, "Putting Cruelty First," *Daedalus* 11, no. 3 (1982): 17.

146 **"relatively light" physical abuse:** This term was used by ANOM: GG AEF: 3H 37: General Prosecutor, AEF Judicial Service, to AEF governor-general, July 11, 1929.

147 **"with the eyes of suffering dogs":** Albert Londres, *Terre d'ébène* (1927; Paris: Collection Motifs, 2006), 258.

147 **"literature, literature!":** Gaston Muraz, *La Vérité sur le chemin de fer Congo-Océan* (Paris: Imprimerie centrale, 1930), 16.

148 **alliances, trade, taxation, intimidation and war:** On politics in the region, see Jan Vansina, *Paths in the Rainforests: Toward a History of Political Tradition in Equatorial Africa* (Madison: University of Wisconsin Press, 1990); and Vansina, "The Peoples of the Forest," in *History of Central Africa,* ed. David Birmingham and Phyllis M. Martin (London: Longman, 1983), 1:75–117.

**148** **"habitual brutality toward the natives":** ANOM: FM 2 AFF POL 19: "Violences commises."

**149** **"mistreatment":** ANOM: FM 2 AFF POL 19: acting lieutenant governor of Ubangi-Shari-Chad to AEF governor-general, August 30, 1913.

**149** **"demand justice from":** ANOM: FM 2 AFFPOL 19: Lieutenant governor of Ubangi-Shari-Chad to general prosecutor in Bangui, *Traduction de la plainte déposée par les arabes: Adam Hssem, Adam Ismail, Ethey Oussmann, Mahmadou, Adboullaye, Abdoullaye Oumr,* April 24, 1913.

**149** **"inadmissible":** ANOM: 2 AFFPOL 19: acting lieutenant governor to AEF governor-general, August 30, 1913; and Inspector Cercus, *Report on the Situation in Lower M'Bomou,* June 30, 1913.

**149** **"an insult to our flag":** ANOM: 2 AFFPOL 19: Teulière to lieutenant governor, July 21, 1913.

**150** **"I will not make them":** ANOM: 2 AFFPOL 19: acting lieutenant governor to AEF governor-general, August 30, 1913; and Cercus, *Report on the Situation in Lower M'Bomou.*

**150** **string a necklace of the ears:** All three of these cases were covered by the newspaper *La Voix des nègres* (January 1927) and were discussed in ANOM: GG AEF: 497: LDH president to minister of colonies, c. November 7, 1927.

**150** **brutality was more common in France:** ANOM: GG AEF: 497: Antonetti to LDH president, January 1928.

**150** **"What worries me the most":** Robert Poulaine, *Étapes africaines: voyage au tour du Congo* (Paris: Nouvelle revue critique, 1930), 47.

**151** **"bagne of the Congo-Océan":** Pierre Contet, in *Le Populaire,* July 26, 1934.

**151** **"social suffering":** See Arthur Kleinman, Veena Das, and Margaret Lock, "Introduction," *Daedalus* 125, no. 1 (1996): 10. (This is a special issue on the subject of social suffering.) For a take on structural violence, see Paul Farmer, "On Suffering and Structural Violence: A View from Below," ibid., 261–83. Pierre Bourdieu refers to a similar phenomenon as "positional suffering" in Bourdieu, ed., *The Weight of the World: Social Suffering in Contemporary Society,* trans. Priscilla Parkhurst Ferguson et al. (Stanford, Calif.: Stanford University Press, 1999), 4.

**151** **"legal rape":** Pierre Contet, in *Le Populaire,* July 22, 1934.

**151** **"croak on the road":** Pierre Contet, in *Le Populaire,* August 2, 1934.

**152** **"are copiously fed":** Ibid.

**152** **"young and pretty girls":** Pierre Contet, in *Le Populaire,* August 11, 1934.

**152** **The boy, named Massima:** "Un scandale en AEF," *La Revue d'outre-mer* 1, no. 9 (November 30, 1932).

**152** **"a shameless sadism":** Pierre Contet, in *Le Populaire,* August 11, 1934.

**153** **Massima's brother watched:** "Un scandale en AEF," *La Revue d'outre-mer* 1, no. 9 (November 30, 1932).

**153** **they were all acquitted:** Pierre Contet, in *Le Populaire,* August 11, 1934. Pro-colonial critics of the decision argued that leaving these "black sheep" unpunished endangered colonial rule in the long run. See, e.g., Jean Philip, "Les Crimes coloniaux," *Les Annales coloniales,* December 24, 1932.

**153** **"too roughly":** ANOM: FM AFFPOL 2856: Labor Service director [Marchand],

*Rapport sur le fonctionnement du Service de la main d'oeuvre de chemin de fer Congo-Océan (de janvier à avril 1925)*, 7–8.

154 **"closed its eyes to certain means"**: Ibid., 20–21.

155 **"an exaggerated emotion"**: Labor Service director, *Rapport sur le fonctionnement (mois de juillet 1925)*, 8.

156 **Colonial authorities quickly grew nostalgic**: ANOM: FM AFF POL 2856: Middle Congo lieutenant governor, "Observations," c. 1925.

156 **"extremely hard on the natives"**: Labor Service director, *Rapport sur le fonctionnement (mois de mai 1925)*, 9–10.

156 **"unqualified to establish authority"**: Labor Service director, *Rapport sur le fonctionnement (mois de juillet 1925)*, 7.

157 **Overseers often lacked gusto**: Ibid., 9.

157 **"painful work"**: Michel R. O. Manot, *L'Aventure de l'or et du Congo-Océan* (Paris: Librairie Secretan, 1946), 48.

157 **"crushed under the vault"**: Labor Service director, *Rapport sur le fonctionnement (mois de juillet 1925)*, 10.

157 **never faced such difficulties**: Manot, *L'Aventure de l'or*, 47.

158 **"Can anyone blame them"**: Labor Service director, *Rapport sur le fonctionnement (mois de juillet 1925)*, 9.

158 **"Mr. Bowen is always hitting us"**: ANOM: FM AFFPOL 2856: *Rapport au sujet de violences dont se serait rendu coupable Monsieur Bowen, tâcheron*, April 30, 1925.

158 **"The men don't want to work"**: Labor Service director, *Rapport sur le fonctionnement (mois de mai 1925)*, 9–10.

159 **"extreme laziness"**: ANOM: FM AFFPOL 2856: *Rapport au sujet de violences dont se serait rendu coupable Monsieur Bowen, tâcheron*, April 30, 1925.

159 **Perrera claimed he did not mean to hurt Tombe**. ANOM: AFFPOL 2856. Hardy de Perini to Labor Service director, April 7, 1925.

160 **"precise facts"**: ANOM: GG AEF: 3H 37: Acting governor-general [de Guise] to Middle Congo lieutenant governor, April 24, 1924.

160 **"no medical conclusion"**: ANOM: AFFPOL 2856: Labor Service director [Thomann] to Labor Service head doctor [Lefrou], May 28, 1925.

160 **"no exterior wounds"**: ANOM: GG AEF: 3H 37: Titaux to Antonetti, April 14, 1928.

160 **another white man, Paoli**: ANOM: GG AEF: 3H 14: General prosecutor to governor-general, June 5, 1926.

160 **men working for Nicolas Caraslanis**: ANOM: GG AEF: 3H 37: *Plaintes de mauvais traitements formulées par les manoeuvres de Monsieur Caraslanis, entrepreneur (CFCO)*, April 25, 1924.

161 **"obvious marks from violent hits"**: ANOM: AFFPOL 2856: Thomann to SCB section head, May 29, 1925.

162 **workers had made unacceptable progress**: ANOM: GG AEF: 3H 36: Antheaume (at Kilometer 71) to SCB division head, September 20, 1927.

162 **"energetically sanction the teams"**: ANOM: GG AEF: 3H 36: Titaux to SCB division head, September 18, 1927.

163 **"You quit working yesterday"**: ANOM: GG AEF: 3H 15: *Procès-verbal de l'enquête*, c. September 16, 1927.

163 **"anarchy spreading across"**: ANOM: GG AEF: 3H 36: SCB chief engineer Girard to Labor Service director, September 19, 1927.

163 **The Frenchman later** *admitted*: ANOM: GG AEF: 3H 36: Girard to Titaux, September 23, 1927.

164 **"a big baton"**: ANOM: GG AEF: 3H 36: Titaux to SCB division head, September 18, 1927.

164 **"all measures you consider useful"**: Ibid.

164 **"extreme brutality"**: ANOM: GG AEF: 3H 36: Titaux to Girard, August 31, 1927.

164 **"must not brutalize the blacks"**: ANOM: GG AEF: 3H 36: Bernet to Boukou Sitou section head, September 3, 1927.

164 **"the natives have shown an indolent"**: ANOM: GG AEF: 3H 36: Bernet (at Kilometer 82) to Girard, September 9, 1927, emphasis added.

164 **"this excellent foreman"**: ANOM: GG AEF: 3H 36: Bernet to Girard, September 14, 1927.

165 **his** *capita,* **Garassena:** He is also called Garassa in certain documents. See, for example, ARC: GG AEF: 348: Antonetti to general prosecutor, March 28, 1928. But in documents throughout the investigation, he is named Garassena; ANOM: GG AEF: 3H 37: Titaux, *Rapport sur l'affaire Garassena,* March 3, 1928.

165 **"a white Italian"**: ANOM: GG AEF: 3H 37: *Procès-verbal d'information,* interrogation of Paul Mboungou, n.d.

165 **An autopsy discovered:** ANOM: GG AEF: 3H 37: Bernet to Titaux, February 23, 1928.

165 **"he was so strong"**: ANOM: GG AEF: 3H 37: Titaux, *Rapport sur l'affaire Garassena,* March 3, 1928.

166 **"disregarding all principle of humanity"**: Ibid.

166 **"Is that all you know?"**: ANOM: GG AEF: 3H 37: *Procès-verbal d'information,* interrogation of Paul Mboungou, n.d. .

166 **"Do you have anything to add?"**: Ibid.

167 **timely fashion required by law:** ANOM: GG AEF: 3H 37: Antonetti to Labor Service director [Titaux], March 28, 1928.

167 **"equally indispensable""**: Ibid.

167 **frequent cases of brutality:** ANOM: GG AEF: 3H 37: Titaux to Antonetti, April 14, 1928.

168 **"indiscipline"**: ANOM: GG AEF: 3H 37: Titaux to governor-general, April 5, 1928.

168 **"We don't hit the blacks"**: Londres, *Terre d'ébène,* 254.

168 **"model detachments"**: Ibid., 265.

169 **a worker named Tchimbana:** ANOM: GG AEF: 3H 37: Prosecutor general, AEF Judicial Service, to governor-general, July 11, 1929.

169 **Caraslanis had previously been accused:** ANOM: GG AEF: 3H 37: *Plaintes de mauvais traitements formulées par les manoeuvres de Monsieur Caraslanis, entrepreneur (CFCO),* April 25, 1924.

169 **One white** *tâcheron:* The incident in mentioned in passing in a file dealing with Chinese laborers. ANOM: GG AEF: 3D 15: Health Service inspector Lasnet, *Rapport sur le début de l'essai de travailleurs chinois au Congo-Océan,* September 20, 1929.

169 **Sexual violence was rarely discussed:** See, for example, the incident in ARC: GG AEF: 361: "Bas-Congo" file, 1932.

170 **"without any motive":** ANOM: GG AEF: 3H 37: Deputy administrator of colonies Lotte to chief administrator of railroad district in Loudima, December 8, 1930.

170 **Roustan's embarassement as a white man:** ANOM: GG AEF: 3H 37: Governor-general to military office head, December 22, 1930.

170 **One worker, N'Gamma:** ANOM: GG AEF: 3H 37: Huet, report no. 16, sector 10; Lt. Pierre-Louisy, response, January 15, 1933; Lt. Pierre-Louisy to Lt. Col. Fevez, Labor Service commander of Congo-Océan Railroad, February 23, 1933; and Lt. Col. Fevez to governor-general, February 27, 1933.

170 **"climate and the ongoing fatigue":** ANOM: GG AEF: 3H 37: Lt. Col. Fevez to governor-general, February 27, 1933.

171 **none of these cases deserved:** ANOM: GG AEF: 3H 37: Prosecutor general, AEF Judicial Service, to governor-general, March 22, 1933.

## CHAPTER 7: DISOBEDIENCE AND DESERTION

172 **everyday interactions between people:** On the everyday uses of means of resistance, see James C. Scott, *Weapons of the Weak: Everyday Forms of Peasant Resistance* (New Haven, Conn.: Yale University Press, 1985), esp. chap. 2.

172 **White overseers oftentimes defended:** See, for example, ANOM: GG AEF: 3H 37: Administrator Lotte to railroad district chief administrator, December 8, 1930; and ANOM: FM AFFPOL 2856: Labor Service director, *Rapport sur le fonctionnement du Service de la main d'oeuvre du chemin de fer Congo-Océan (mois de juillet 1925)*, August 16, 1925, 8.

173 **images of bridges and rails:** ANMT: 89 AQ 1000: Photographies.

174 **"the signs of a profound interior rebellion":** Labor Service director, *Rapport sur le fonctionnement (mois de juillet 1925)*.

175 **"sinister rumors":** ANOM: GG AEF: 3D 15: Health Service inspector Lasnet, *Rapport sur le début de l'essai de travailleurs chinois au Congo-Océan*, September 20, 1929.

175 **The Chinese government had officially:** On Chinese workers in the Congo, see Julia Martínez, "'Unwanted Scraps' or 'An Alert, Resolute, Resentful People'?: Chinese Railroad Workers in French Congo," *International Labor and Working-Class History* 91 (Spring 2017): 79–98.

175 **786 Chinese workers:** ANOM: GG AEF: 3D 15: Health Service inspector Lasnet, *Rapport sur le début de l'essai de travailleurs chinois au Congo-Océan*, September 20, 1929.

175 **"muscular and robust":** Ibid.

176 **"appetizing and properly served":** Ibid.

176 **The Chinese were deemed ineffective:** ANOM: GG AEF: 3H 48: *Les Travailleurs chinois en AEF,* September 23, 1929.

176 **In South Africa in the early 1900s:** Gary Kynoch, "Controlling the Coolies: Chinese Mineworkers and the Struggle for Labor in South Africa, 1904–1910," *International Journal of African Historical Studies* 36, no. 2 (2003): 309–29.

177 **They backed down only:** For detailed accounts from Captain Houdré, see ANOM: GG AEF: 3H 48; and ANOM: GG AEF: 3D 15: Lasnet, *Rapport sur le début de l'essai de travailleurs chinois.*

177 **"bad dispositions":** ANOM: GG AEF: 3D 15: Lasnet, *Rapport sur le début de l'essai de travailleurs chinois.*

177 **"insufficiency of repression":** Ibid.

178 **"strangely resembled Communist methods":** ANOM: GG AEF: 3H 48: Labor Service director to governor-general, August 14, 1929. The Chinese Communist Party was founded in 1921.

178 **"sufficient repressive solution":** ANOM: GG AEF: 3D 15: Lasnet, *Rapport sur le début de l'essai de travailleurs chinois.*

178 **172 other workers were deported:** Martínez, "'Unwanted Scraps,'" 92.

178 **five Chinese workers had already died:** ANOM: GG AEF: 3D 15: Lasnet, *Rapport sur le début de l'essai de travailleurs chinois.*

178 **workers in the highlands demanded:** ANOM: GG AEF: 3H 48: Lt. Col. Allut (M'Boulou) to acting governor-general, February 13, 1930.

178 **"threatened progress":** ANOM: GG AEF: 3H 48: Rouberol to Antonetti, February 11, 1930.

179 **"I ask only to go":** ANOM: GG AEF: 3H 48: Lt. Col. Allut to acting governor-general, February 13, 1930.

179 **With contracts in hand:** ANOM: GG AEF: 3H 48: Labor Service director to governor-general, August 14, 1929.

179 **fresh fish and chicken, and even opium:** ANOM: GG AEF: 3H 48: Captain Houdré to Labor Service director, August 13, 1929.

179 **forcing Chinese prisoners to work without pay:** Martínez, "'Unwanted Scraps,'" 92.

180 **two Chinese were apparently murdered:** Rita Headrick, *Colonialism, Health and Illness in French Equatorial Africa, 1885–1935* (Atlanta, GA.: African Studies Association Press, 1994), 304.

181 **workers would leave their worksites to file complaints:** ANOM: GG AEF: 3H 37: Middle Congo acting lieutenant governor to Mindouli subdivision head, February 26, 1924.

181 **"mutinies":** Mario Azevedo, "The Human Price of Development: The Brazzaville Railroad and the Sara of Chad," *African Studies Review* 24, no. 1 (1981): 7.

181 **recruit 5 percent above the target quota:** ANOM: AFFPOL 2856: Labor Service director Thomann to Middle Congo lieutenant governor, February 14, 1925.

182 **two *miliciens* who departed:** ANOM: GG AEF: 3H 14: Dubois, "Note sur la situation actuelle au B.O.," July or August 1925.

182 **of the 327 men registered:** ANOM: GG AEF: 3H 9: Health Service inspector [Lasnet], *Mission de l'Inspection,* March 1, 1928.

182 **a contingent of 159 recruits:** ANOM: GG AEF: 3H 37: Head administrator and inspector of administrative affairs to governor-general, January 2, 1929.

182 **officials linked brutal European oversight:** ANOM: AFFPOL 2856: Labor Service director Thomann to Middle Congo lieutenant governor, May 13, 1925, 8.

183 **replaced by another from his home village:** ANOM: GG AEF: 3H 37: Marchand to Mindouli subdivision head, February 26, 1924.

183 **"the most effective means":** ANOM: GG AEF: 3H 37: Marchand to railroad district head in Loudima, July 28, 1924.

183 **"exemplary punishments":** ANOM: GG AEF: 3H 37: Principal engineer for public works 2nd class to Middle Congo lieutenant governor, July 30, 1925.

183 **reached a ratio of one in four:** ANOM: GG AEF: 3H 37: Marchand to railroad district head in Loudima, July 28, 1924.

183 **forty-three workers deserted:** ANOM: GG AEF: 3H 37: Engineer Nicolau to Middle Congo lieutenant governor, August 6, 1925.

183 **more than 300 of the 511 workers:** Joseph Gamandzori, "Chemin de fer: villes et travail au Congo (1921–1953)," Ph.D. diss., University of Lille (1987), 195.

184 **"excessive indolence":** ANOM: GG AEF: 3H 37: Marchand to Mindouli subdivision head, February 26, 1924.

184 **"sickly" due to *nostalgie*:** On the medical history of nostalgia and homesickness, see Thomas Dodman, *What Nostalgia Was: War, Empire, and the Time of a Deadly Emotion* (Chicago: University of Chicago Press, 2018).

184 **"The workers in any case":** ANOM: GG AEF: 3H 37: Chief administrator [Marchessou] to governor-general, January 2, 1929.

184 **The majority of deserters:** Gilles Sautter, "Notes sur la construction du chemin de fer Congo-Océan (1921–1934)," *Cahiers d'études africaines* 7, no. 26 (1967): 249.

184 **lost between 6 and 25 percent:** Ibid.

184 **pretended to be dead:** ANOM: GG AEF: 3H 7: Inspector of administrative affairs to governor-general, February 12, 1929.

185 **"The spectacle was beautiful":** Pierre Contet, in *Le Populaire*, July 26, 1934.

186 **"thin, emaciated to the bone":** Ibid.

186 **"We come from M'Vouti":** Ibid.

187 **only the slimmest hope of defying:** Ibid.

187 **"contempt for human life":** ibid.

187 **"primitives":** ANOM: GG AEF: 3H 37: Marchand to Mindouli subdivision head, February 26, 1924.


## CHAPTER 8: THE MANY WAYS OF DEATH

188 **"breathing very well":** ANOM: GG AEF: 3H 28: Marichelle to delegate, September 23, 1926.

188 **failed to either corroborate or disprove:** ANOM: GG AEF: 3H 28: Administrative delegate of general government and of Middle Congo to Antonetti, October 9, 1926.

189 **"relative humanity":** ANOM: GG AEF: 3H 8: Pégourier to minister of colonies, October 15, 1926.

189 **"all measures likely to minimize":** ANOM: GG AEF: 3H 8: Boyé, *Rapport médical annuel (octobre 1924–décembre 1925)*, 114–15.

189 **estimated that 22.5 percent of workers:** ANOM: GG AEF: 3H 8: Boyé to governor-general, August 11, 1925.

190 **manifestations of the ineffectiveness:** The Congo-Océan was not the only place in the French Empire where the state grappled with such issues. On the battle against famine across the colonies, see Yan Slobodkin, "Empire of Hunger: Famine and the French Colonial State, 1867–1945," Ph.D. diss., Stanford University (2018).

191 **"lamentable conditions of this":** ANOM: GG AEF: 3H 14: Du Bois, "Note sur la situation actuelle au B.O.," July or August 1925.

191 **Workers, he pointed out:** Albert Londres, *Terre d'ébène* (1927; Paris: Collection Motifs, 2006), 245.

191 **there was little food:** ANOM: GG AEF: 3H 14: Du Bois, "Note sur la situation actuelle."

192 **white agents needed 150 francs:** ANMT: 89 AQ 873: Rouberol to Augagneur, February 15, 1923.

192 **"emaciated, skeletal":** Georges Lefrou, "Contribution à l'étude de l'utilisation de la main-d'oeuvre indigène," *Annales de médecine et de pharmacie coloniales,* no. 25 (1927): 31.

192 **recruits were supplied with all they needed:** ANOM: GG AEF: 3H 8: Acting lieutenant govenor, "Observations" (c. August 1925). This report responds to the letter from health inspector [Boyé] to governor-general, August 11, 1925.

192 **Others sold or gave away their supplies:** Rita Headrick, *Colonialism, Health and Illness in French Equatorial Africa, 1885–1935* (Atlanta: African Studies Association Press, 1994), 280.

192 **Dispensing cash rather than food:** ANOM: GG AEF: 3H 7: Prouteaux to governor-general, January 11, 1927.

193 **staples varied:** On food production and consumption, see Headrick, *Colonialism, Health and Illness,* 9–13. On workers' desired food, see also ANOM: GG AEF: 3H 8: Boyé, *Rapport medical annuel (octobre 1924–décembre 1925),* 121.

193 **"famished wretched":** ANOM: GG AEF: 3H 7: Levet to head of Bureau of Political Affairs of General Government, May 1, 1924.

194 **A rice ration was set:** Dr. Georges Lefrou, who was the main health inspector in 1925, claimed the rice ration was 800 grams per day; by contrast, many documents stated it as 500 grams, which itself was not always delivered. See Lefrou, "Contribution à l'étude de l'utilisation," 30, 32.

194 **"a heavy task":** Gilles Sautter, "Notes sur la construction du chemin de fer Congo-Océan (1921–1934)," *Cahiers d'études africaines* 7, no. 26 (1967): 265.

194 **"As for crops":** ANOM: AFFPOL 2856: Labor Service director [Thomann] to Middle Congo lieutenant governor, May 13, 1925.

194 **palm oil, a common ingredient:** On shifting rations, see Headrick, *Colonialism, Health and Illness,* 283.

195 **fourteen cows were slaughtered:** ANOM: GG AEF: 3H 8: Boyé, *Rapport médical annuel (octobre 1924–décembre 1925),* 119.

195 **beef was welcomed:** ANOM: GG AEF: 3H 7: Houillon, *Extrait du rapport d'inspection du médecin principal* (c. 1926–27).

195 **"all this represents the elements":** ANOM: GG AEF: 3H 8: Boyé, *Rapport médical annuel (octobre 1924–décembre 1925),* 118–21.

196 **Some 15,000 kilos were needed:** ANOM: GG AEF: 3H 37: Public works inspector [Marchand] to governor-general, January 28, 1924.

196 **loss of foodstuffs in transport:** ANOM: GG AEF: 3H 7: Levet to head of Bureau of Political Affairs of General Government, May 1, 1924.

196 **how to make manioc flour:** Lefrou, "Contribution à l'étude de l'utilisation," 36–37.

196 **"all the parasites, friends":** Ibid.

196 **to steal from one another and to raid:** ANOM: FM AFFPOL 2856: Labor Service director, *Rapport sur le fonctionnement du Service de la main d'oeuvre du chemin de fer (mois de Juillet 1925),* August 16, 1925.

197 **"threaten not only to quit":** ANMT: 89 AQ 873: Rouberol to governor-general, November 7, 1925.

197 **"Even our starved skeletal invalids":** ANOM: GG AEF: 3H 8: [LeFrou,] *Chemin*

*de fer Brazzaville-Océan, division côtière, ambulance de Mavouadi, compte rendu mensuel, année 1925, mois de juin.*

197 **"What rot!":** Marcel Homet, *Congo: terre de souffrances* (Paris: Fernand Aubier, 1934), 243.

197 **"And in accordance with these":** Marcel Homet, "La Vérité sur l'AEF," *Esprit: revue internationale* 2, no. 18 (1934): 919.

198 **"The experience of suffering":** Paul Farmer, "On Suffering and Structural Violence: A View from Below," *Race/Ethnicity: Multidisciplinary Global Contexts* 3, no. 1 (2009): 12; originally published in *Daedalus* 125, no. 1 (1996): 251–83.

198 **"very poor":** Lefrou, "Contribution à l'étude de l'utilisation," 33.

199 **sustenance provided in concentration camps:** For example, on diets in occupied Germany after the Second World War, see Atina Grossmann, "Grams, Calories, and Food: Languages of Victimization, Entitlement, and Human Rights in Occupied Germany, 1945–1949," *Central European History* 44, no. 1 (2011): 118–48.

199 **ILO set the standard caloric intake:** Nick Cullather, "The Foreign Policy of the Calorie," *American Historical Review* 112, no. 2 (2007): 337–64; discussion of the ILO is on 355.

199 **"the response was invariably":** ANOM: GG AEF: 3H 7: Levet to head of Bureau of Political Affairs of General Government, May 1, 1924.

200 **inconsistent diet of European workers:** ANOM: FM AFFPOL 2856: Labor Service director, *Rapport sur le fonctionnement du (mois de juillet 1925).*

200 **read like Parisian shopping lists:** ANMT: 89 AQ 946: "Lettres de Rouberol," no. 1, List of supplies ordered, January 9, 1923.

200 **"non-obese individuals":** Claude Piantadosi, *The Biology of Human Survival: Life and Death in Extreme Environments* (New York: Oxford University Press, 2003), 35.

201 **effects of prolonged semistarvation:** John R. Butterly and Jack Shepherd, *Hunger: The Biology and Politics of Starvation* (Hanover, N.H.: Dartmouth College Press, 2010), 56.

201 **"Minnesota experiment":** The findings of the Minnesota experiment were published as Ancel Keys et al., *The Biology of Human Starvation*, 2 vols. (Minneapolis: University of Minnesota Press, 1950).

201 **Severe hunger renders people:** The interest of the Minnesota experiment to sociologists and psychologists was unmistakable. See, for example, W. F. Ogburn's review in the *American Journal of Sociology* 57, no. 3 (1951): 294–95; David Baker and Natacha Keramidas, "The Psychology of Hunger," *Monitor on Psychology* 44, no. 9 (2013): 66; and Butterly and Shepherd, *Hunger,* 57–58.

202 **failure "to adapt":** ANOM: GG AEF: 3H 8: Boyé, *Rapport médical annuel (octobre 1924–décembre 1925),* 123.

202 **hunger stared at them gauntly:** Richard Wright, *Black Boy* (1945; New York: HarperPerennial, 2007), 14.

203 **"remarkable state of emaciation":** ANOM: GG AEF: 3H 8: Pégourier to minister of colonies, October 15, 1926.

203 **"the convoys wait in vain":** Albert Londres, *Terre d'ébène* (1927; Paris: Collection Motifs, 2006), 245.

203 **"Tortured by hunger":** Headrick, *Colonialism, Health and Illness,* 284.

204 **beriberi is caused by a vitamin B1:** On the discovery of the causes of beriberi, see

Yan Slobodkin, "Famine and the Science of Food in the French Empire, 1900–1939," *French Politics, Culture and Society* 36, no. 1 (2018): 52–75, esp. 55–57.

204 **"undetermined":** ANOM: GG AEF: 3H 8: Boyé, *Rapport médical annuel (octobre 1924–décembre 1925),* 116.

204 **weakness and generalized edema:** Lefrou, "Contribution à l'étude de l'utilisation," 31.

204 *misère physiologique:* ANOM: GG AEF: 3H 7: Houillon, *Extrait du rapport du médecin principal,* report no. 1H, *Mission Pégourier,* c. 1927.

205 **beriberi and malnutrition accounted for about 13 percent:** Lefrou, "Contribution à l'étude de l'utilisation," 50.

205 **"strictly administrative nature":** ANOM: GG AEF: 3H 8: Boyé, *Rapport médical annuel (octobre 1924–décembre 1925),* 116.

206 **bacterial infections that workers faced:** Claude A. Piantadosi, *The Biology of Human Survival: Life and Death in Extreme Environments* (Oxford: Oxford University Press, 2003), 35.

206 **Dangers on the construction site:** ANOM: GG AEF: 3H 8: *Chemin de fer Brazzaville-Océan, division côtière, ambulance de Mavouadi, compte rendu mensuel* (May 1925).

206 **A yellow fever epidemic in Matadi:** ANOM: GG AEF: 3H 34: Governor secretary general of AEF General Government, report, November 21, 1928.

206 **500 cases of respiratory infections:** Headrick, *Colonialism, Health, and Illness,* 292.

206 **African trypanosomiasis:** For a comparative colonial take on sleeping sickness, see Maryinez Lyons, *The Colonial Disease: A Social History of Sleeping Sickness in Northern Zaire, 1900–1940* (Cambridge, U.K.: Cambridge University Press, 1992).

206 **"most murderous":** Lefrou, "Contribution à l'étude de l'utilisation," 48.

207 **15 percent suffered from the disease:** Headrick, *Colonialism, Health, and Illness,* 285.

207 **"thick streaks or more or less":** Adolphe Nicolas et al., *Guide hygiénique et médical du voyageur dans l'Afrique centrale* (Paris: Challamel Aîné, 1885), 429–31.

207 **calomel, a toxic purgative:** ANOM: GG AEF: 3H 8: *Chemin de fer Brazzaville-Océan, division côtière, ambulance de Mavouadi, compte rendu mensuel* (May 1925).

208 **"bland and nauseous":** Nicolas, *Guide hygiénique,* 430–31.

208 **"preferring to defecate around the hut":** ANOM: GG AEF: 3H 8: *Chemin de fer Brazzaville-Océan, division côtière, ambulance de Mavouadi, compte rendu mensuel* (May 1925).

208 **one official encouraged the construction of isolation wards:** Nicolas, *Guide hygiénique,* 430–31.

208 **"a superstitious terror":** Labor Service director, *Rapport sur le fonctionnement (mois de juillet 1925),* 15. On colonial medicine as a field, see Helen Tilley, "Medicine, Empires, and Ethics in Colonial Africa," *AMA Journal of Ethics* 18, no. 7 (2016): 743–53.

209 **the origin and histories of the various:** A useful overview of competing theories is Jim Moore, "The Puzzling Origins of AIDS," *American Scientist* 92 (2004): 540–47.

209 **The story, with its globe-trotting scientists:** One engaging scientific account is David Quammen, *The Chimp and the River: How AIDS Emerged from an African Forest* (New York: W.W. Norton, 2015).

209 **Disagreements about the origin of HIV-1:** Consider, for example, the debates around E. Hooper, *The River: A Journey to the Source of HIV and AIDS* (Boston: Little Brown, 1999).

209 *Pan troglodytes troglodytes*: Jim Moore, "The Puzzling Origin of AIDS," *American Scientist* 92 (2004), 540–47; and Jacques Pepin, *The Origin of AIDS* (Cambridge, U.K.: Cambridge University Press, 2011), chap. 4.

210 **this spillover occurred, at the latest:** Brandon F. Keele, "Chimpanzee Reservoirs of Pandemic and Non-pandemic HIV-1," *Science* 313 (July 28, 2006): 523–26; and Michael Worobey, "Direct Evidence of Extensive Diversity of HIV-1 in Kinshasa by 1960," *Nature* 455 (October 2, 2008): 661–65.

210 **From there, the virus moved south:** Nuno Faria et al., "The Early Spread and Epidemic Ignition of HIV-1 in Human Populations," *Science* 346, no. 6205 (October 3, 2014): 56–61.

210 **conducting autopsies on Congo-Océan workers:** Pepin, *Origins of AIDS*, 36–37.

210 **"cachexia of unknown origin":** Léon Pales, "La tuberculose des noirs vue d'Afrique équatoriale française," *Revue de la tuberculose* 5, no. 4 (1938): 202.

211 **"Mayombe cachexia":** Pepin, *Origins of AIDS*, 37–38.

211 **chronic nonbloody diarrhea:** Ibid.

211 **physical and psychological misery:** For a detailed description of symptoms in patients in the 1980s, see Bradford A. Navia et al., "The AIDS Dementia Complex: I. Clinical Features," *Annals of Neurology* 19, no. 6 (1986): 517–24.

212 **"this does not exclude anything":** Pepin, *Origins of AIDS*, 38.

212 **"we will never know for sure":** Ibid., 39.

212 **the disease had been locally present:** Michael Worobey, "1970s and 'Patient 0' HIBV-1 Genomes Illuminate Early HIV/AIDS History in North America," *Nature* 539, no. 7627 (2016): 98–101.

213 **early transmission were between 1909:** Faria et al., "Early Spread and Epidemic Ignition," 57.


CHAPTER 9: A BUREAUCRAT'S HUMANITARIANISM

214 **"It is incontestable that big mistakes":** ANOM: FM AFFPOL 2856: Labor Service director, *Rapport sur le fonctionnement du Service de la main d'oeuvre du chemin de fer Congo-Océan (mois de juillet 1925),* August 16, 1925, emphasis added.

215 **"pulmonary fragility":** ANOM: GG AEF: 3H 8: Boyé, *Rapport médical annuel (octobre 1924–décembre 1925),* emphasis in original.

215 **"all the precautions":** Ibid.

216 **only eighteen doctors in all:** Rita Headrick, *Colonialism, Health and Illness in French Equatorial Africa, 1885–1935* (Atlanta: African Studies Association Press, 1994), 228.

216 **twenty-nine doctors in the entire:** ANOM: GG AEF: 3H 9: Antonetti to minister of colonies, June 17, 1928.

216 **Many physicians who came from France:** Headrick, *Colonialism, Health and Illness,* 231.

216 **preferred traditional healers:** Deborah Neill, *Networks in Tropical Medicine: Internationalism, Colonialism, and the Rise of a Medical Specialty, 1890–1930* (Stanford, Calif.: Stanford University Press, 2012), chap. 3.

217 **Historians of imperialism have debated:** A sizable scholarship on this theme exists; but examples of the two camps can be found by comparing Paul Rabinow, *French Modern: Norms and Forms of the Social Environment* (Chicago: University

of Chicago Press, 1995), and Peter Zinoman, *The Colonial Bastille: A History of Imprisonment in Vietnam, 1860–1940* (Berkeley: University of California Press, 2001). On the modern state more generally, see James Scott, *Seeing Like a State: How Certain Schemes to Improve the Human Condition Have Failed* (New Haven, Conn.:Yale University Press, 1998).

217 **a common standard for what makes:** See, for example, Michel Foucault, *The History of Sexuality,* vol. 1, *An Introduction,* trans. Robert Hurley (New York:Vintage, 1978), 138.

219 **"an energy without equal":** Jean Weber, "Le Gouverneur Général Antonetti," *La Revue des ambassades et des questions diplomatiques et coloniales* (May 1938): 14.

219 **"a man of character, in a":** Marcel Sauvage, *Les Secrets de l'Afrique noire* (Paris: Denoël, 1937), 67–68.

219 **"quick to make decisions":** Albert Londres, *Terre d'ébène* (1927; Paris: Collection Motifs, 2006), 240.

220 **the modern French state was built on paperwork:** See, for example, Ben Kafka, *The Demon of Writing: The Powers and Failures of Paperwork* (Princeton: Princeton University Press, 2012).

220 **"All the staffs on which":** Marc Bloch, *Strange Defeat: A Statement of Evidence Written in 1940,* trans. Gerard Hopkins (NewYork:W.W. Norton, 1999), 58.

221 **inspections and investigations:** Catherine Cocquery-Vidrovitch, *Le Congo au temps des grands compagnies concessionnaires, 1898–1930* (Paris: Mouton, 1972), 90.

221 **asked to investigate themselves:** Some areas of West Africa were not visited by inspectors until the 1930s. William B. Cohen, *Rulers of Empire: The French Colonial Service in Africa* (Stanford, Calif.: Hoover Institute Press, 1971), 64.

221 **"with the collaboration of local officials":** ALDH: F Delta rés 798/89: LDH president to minister of colonies, February 1, 1933.

221 **"Parliaments are certainly kept":** AILO: N206/1/88/1: Raymond Buell (in Belgian Congo) to Grimshaw, January 29, 1926.

222 **techniques to disavow and reinterpret accusations:** On the varieties of denial, see Stanley Cohen, *States of Denial: Knowing About Atrocities and Suffering* (Malden, Mass.: Polity Press, 2001).

223 **"I can no longer keep myself":** ANOM: AFFPOL 2856: Minister of colonies to Antonetti, August 5, 1928.

223 **"abuse of a native":** ANOM: AFFPOL 2856: Antonetti to minister of colonies, April 5, 1928.

223 **Maran maintained that he was innocent:** Alice Conklin, "Who Speaks for Africa? The René Maran–Blaise Diagne Trial in 1920s Paris," in *The Color of Liberty: Histories of Race in France,* ed. Sue Peabody and Tyler Stovall (Durham, N.C.: Duke University Press, 2003), 308–9.

223 **"tendentious campaigns":** ANOM: AFFPOL 2856: Antonetti to minister of colonies, June 8, 1929.

224 **dense, eighteen-page typed rebuttal:** ANOM: AFFPOL 2856: Antonetti to minister of colonies, October 28, 1925.

224 **level-headedness and reason must rule:** ANOM: AFFPOL 2856: Antonetti to minister of colonies, October 28, 1925. In this way, Antonetti was not out of step with what William M. Reddy calls the "emotional regime" of the age where "sentimentalism"

was acceptable in literature but not in the cold world of administration. See Reddy, *The Navigation of Feeling: A Framework for the History of Emotions* (Cambridge, U.K.: Cambridge University Press, 2001), esp. chap. 7.

224 **"respectfully but strongly"**: ANOM: AFFPOL 2856: Antonetti to minister of colonies, October 28, 1925.

225 **"If one had bothered to photograph"**: Ibid.

225 **"racial shaming"**: David Leverenz, *Honor Bound: Race and Shame in America* (Piscataway, N.J.: Rutgers University Press, 2012), esp. chap. 1.

225 **sympathize with white men or Africans:** For an examination of how group identities gave rise to "emotional communities," see Barbara H. Rosenwein, *Emotional Communities in the Early Middle Ages* (Ithaca, N.Y.: Cornell University Press, 2006), esp. 23–29.

226 **"treated fairly"**: ANOM: GG AEF: 3H 8: Report no. 8 of colonial inspector Sol, no date.

226 **"shows how all these statistics"**: Ibid.

227 **"an outlet to the huge"**: *L'Echo de Paris,* September 24, 1924.

227 **"everyone in the Congo"**: ANOM: AFFPOL 2856: Antonetti to minister of colonies, October 28, 1925.

227 **"there reigns in this country"**: ANOM: GG AEF: 3H 8: Responses of the governor general to the Mission Pégourier, c. 1926.

228 **"as soon as I arrived"**: Ibid.

228 **"humanity forces us, *alas!*"**: ANOM: AFFPOL 2856: M. R. Antonetti, *Discours prononcé à la séance d'ouverture du Conseil du Gouvernement, session ordinaire de décembre 1925* (Brazzaville: Gouvernement général, 1926), 7.

228 **"my doctrine"**: ANOM: GG AEF: 3H 8: Responses of the governor general to the Mission Pégourier, c. 1926.

228 **"legible"**: On legibility and the state building, see James C. Scott, *Seeing Like a State: How Certain Schemes to Improve the Human Condition Have Failed* (New Haven, Conn.: Yale University Press, 1998); and Akhil Gupta, *Red Tape: Bureaucracy, Structural Violence, and Poverty in India* (Durham, N.C.: Duke University Press, 2012).

228 **Reports informed regulations:** On the relationship between sovereignty and the power to determine who lives and dies, see Achille Mbembe, "Necropolitics," *Public Culture* 15, no. 1 (2003): 11–40. For a useful overview of different approaches to bio-power, see Thomas Lemke, *Bio-politics: An Advanced Introduction* (New York: NYU Press, 2011).

229 **embodied the very ideology, concerns:** On the relationship between bureaucracy and paperwork, see Matthew S. Hull, *Government of Paper: The Materiality of Bureaucracy in Urban Pakistan* (Berkeley: University of California Press, 2012), 1–33.

229 **"minute precision of military orders"**: ANOM: AFFPOL 2856: Antonetti to minister of colonies, October 28, 1925.

229 **"the care"**: Ibid.

230 **"in a fashion that can offer"**: Ibid.

231 **a five-fold increase:** These numbers are based on Sautter's estimates. Gilles Sautter, "Notes sur la construction du chemin de fer Congo-Océan (1921–1934)," *Cahiers d'études africaines* 7, no. 26 (1967): 271.

231 **"It is profoundly saddening"**: ANOM: FM AFFPOL 2856: Military inspector,

"Note personnelle pour le ministre au sujet de l'organisation de la main d'oeuvre du chemin de fer Congo-Océan," Paris, June 4, 1928.

232 **many did not live to fulfill their contracts:** ANOM: GG AEF: 3H 34: *Rapport du gouverneur secrétaire général du gouvernement général de l'AEF,* November 21, 1928.

232 **expected attrition due to illness, death:** Sautter, "Notes sur la construction," 240.

232 **In an interview from the 1970s:** Mario Azevedo, "The Human Price of Development: The Brazzaville Railroad and the Sara of Chad," *African Studies Review* 24, no. 1 (1981): 7.

233 **"dispelled boredom":** Georges-R. Manue, "Sur les chantiers du Congo-Océan," *La Dépêche coloniale,* October 16, 1930.

233 **"excellent effect":** ANOM: GG AEF: 3H 34: *Rapport du gouverneur secrétaire général du gouvernement général de l'AEF,* November 21, 1928.

233 **mortality rates represented comparative:** Sautter, "Notes sur la construction," 271.

233 **beriberi was still a problem into the 1930s:** ANOM: GG AEF: 3H 58: Lecomte, report, February 9–21, 1931.

234 **white overseers cheating workers:** ANOM: GG AEF: 3H 34: Congo-Océan subdivision head to SCB head, May 21, 1931.

234 **more than fifteen hundred men agreed:** ANOM: GG AEF: 3H 58: Lecomte, report, February 9–21, 1931.

234 **"I did everything":** ANOM: GG AEF: 3H 7: Antonetti to Lasnet, August 24, 1929.

235 **"spirit of achievement":** Ibid.

235 **"the one true solution":** ANOM: AFFPOL 2856: Antonetti to minister of colonies, June 8, 1929.

235 **new sources of revenue that the railroad:** ANOM: GG AEF: 5D 47: Antonetti to minister of colonies, June 9, 1929.

235 **"under the attentive and":** ANOM: GG AEF: 3H 58: Alfassa to Antonetti, June 3, 1931.

236 **only 0.33 percent of the population:** ANOM: GG AEF: 5D 47: Antonetti to minister of colonies, June 8, 1929.

236 **"The accepted sacrifices":** Ibid.

236 **"total care":** On the failures of "total care" in prison conditions, see Peter Zinoman, *The Colonial Bastille: A History of Prisons and Imprisonment in Vietnam, 1862–1940* (Berkeley: University of California Press, 2001), esp. chap. 3. On the Indochinese plantations, see Michitake Aso, *Rubber and the Making of Vietnam: An Ecological History, 1897–1975* (Chapel Hill: University of North Carolina Press, 2018). See also "La grande pitié des travailleurs annamites," written by "M.D.," an anonymous colonial inspector, which appeared in *La Résurrection* in 1928, reprinted in Félicien Challaye, *Un livre noir du colonialisme: "Souvenirs sur la colonisation"* (Millau: Nuits Rouges, 2003), 155–65.

237 **The maps drawn, data collected:** They were not unlike the "panoramas" of the social described by Bruno Latour in *Reassembling the Social: An Introduction to Actor-Network-Theory* (Oxford: Oxford University Press, 2005).

237 **"There's not a colonial project":** Pierre Mille, "La 'bataille' du chemin de fer," *La Dépêche coloniale,* March 9, 1927.

## CHAPTER 10: SILENCING CRITICS

**240** **"half-baked antinomian opinion":** George Orwell, *The Road to Wigan Pier* (New York: Harcourt, Brace, 1958), 138.

**240** **creation of international bodies:** The historiography of humanitarianism is growing daily. On empire and humanitarianism in the British case, see Emily Baughan and Bronwen Everill, "Empire and Humanitarianism: A Preface," *Journal of Imperial and Commonwealth History* 40, no. 5 (2012): 727–28. On the renewed interest in the League of Nations, see Susan Pedersen, *The Guardians: The League of Nations and the Crisis of Empire* (New York: Oxford University Press, 2015); and Pedersen, "Back to the League of Nations," *American Historical Review* 112, no. 4 (2007): 1091–117.

**240** **representatives from over thirty countries and colonies:** Archives of the Internationaal Instituut voor Sociale Geschiedenis (Amsterdam): 33921: League Against Imperialism archives: List of Organizations and Delegates Attending the Congress Against Colonial Oppression and Imperialism, Brussels, February 10, 1927.

**241** **"to look over the screen":** Albert Londres, *Terre d'ébène* (1927; Paris: Collection Motifs, 2006), 271, 268–69.

**241** **"suffering":** Marcel Homet, *Congo: terre de souffrances* (Paris: Fernand Aubier, 1934), preface.

**241** **"a little humanity":** Marcel Sauvage, *Les Secrets de l'Afrique noire* (Paris: Denoël, 1937), 332.

**241** **"a humane and rational":** ANOM: AFFPOL 2856: René Maran to minister of colonies, January 26, 1928. This claim was corroborated in Londres, *Terre d'ébène,* 243–44.

**241** **"to ensure respect for":** Victor Augagneur, "La Grande Misère des noirs du Congo," *Oeuvre,* June 10, 1927.

**242** **"To think that de Brazza":** Pierre Contet, in *Le Populaire,* August 12, 1934.

**242** **"Brazza wasn't there":** Londres, *Terre d'ébène,* 269.

**242** **"doctrinal firmness and his rigid honesty":** "Simple question," *Bec et Ongles* (December 31, 1932).

**242** **"inhumane":** ANOM: GG AEF: 3H 8: Pégourier to minister of colonies, October 15, 1926.

**242** **"bad treatment":** See Jürgen Zimmerer and Joachim Zeller, *Genocide in German South-West Africa: The Colonial War (1904–1908) in Namibia and Its Aftermath,* trans. Edward Neather (Monmouth, U.K.: Merlin Press, 2008).

**244** **"on each of these":** Camille Guy, "L'Avenir de l'Afrique équatoriale française," *Bulletin de l'Afrique française* (January 1926): 9, 13.

**244** **"Transit routes, hygiene":** Charles Debierre, "Quelle est la véritable population de l'Afrique équatoriale française?," *Les Annales coloniales,* September 10, 1926.

**244** **expressed faith in the regulations:** See, for example, "Sur les chantiers de Brazzaville," *Les Annales coloniales,* January 17, 1927.

**244** **"inexhaustible quantities":** Anselme Laurence, "L'Afrique équatoriale française" *Journée industrielle* (May 29, 1926).

**245** **"food problem":** Georges Boussenot, "Le Congo-Océan est nécessaire à la vie de l'AEF," *Bulletin de l'Afrique française* (August 1926): 400.

**246** **"lies and calumnies":** "Les Mensonges et les calomnies contre l'Afrique équatoriale française," *Bulletin de l'Afrique française* (February 1929): 57–59.

246 **"legends":** Bernard Desouches, "La Question du chemin de fer Congo-Océan," *Illustration,* no. 4505 (July 1929). See also, for example, Maurice Rondet-Saint, *Sur les routes du Cameroun et de l'AEF* (Paris: Société d'éditions géographiques, maritimes et coloniales, 1933), 177.

246 **"vast necropolis of blacks":** Louis Proust, "La Vérité sur le Brazzaville-Océan," *La Dépêche coloniale,* July 3, 1929.

247 **"In Africa, we must innovate":** Julien Maigret, "Le Chemin de fer Brazzaville-Océan," *La Chronique coloniale,* February 15, 1929.

247 **"gigantic work that still":** Gaston Muraz, *La Vérité sur le chemin de fer Congo-Océan* (Paris: Imprimerie centrale, 1930), 8–9, 17.

248 **"You cannot generalize by speaking":** Henri Fontanier and Léon Perrier in *Journal officiel de la République française. Débats parlementaires. Chambre des députés,* November 23, 1927, 3179.

248 **"When one wants to know the truth":** André Maginot, ibid., June 14, 1929, 2061.

249 **"Forced labor is disguised slavery":** Nouelle quoted in "Á la Chambre," *Les Annales coloniales,* June 15, 1929.

249 **"Let me be understood":** Henry Fontanier in *Journal officiel de la République française. Débats parlementaires. Chambre des députés,* November 23, 1927, 3178.

249 **Few called for major changes:** Martin Thomas, *The French Empire between the Wars: Imperialism, Politics, and Society* (Manchester, U.K.: Manchester University Press, 2005), esp. chap. 1.

250 **"slavery in all its forms":** This point was made explicit by ALN: Mandates, R66/45242/23252, Sir Frederick Lugard. "Note sur le travail forcé," July 29, 1925.

250 **to draft a forced labor convention:** For a succinct overview, see Joseph P. Chamberlain, "Forced Labor," *Annals of the American Academy of Political and Social Science* 166, no. 1 (1933): 80–85; and Adom Getachew, *Worldmaking After Empire: The Rise and Fall of Self-Determination* (Princeton: Princeton University Press, 2019), esp. chap. 2.

251 **"to take in hand the control of our Colonies":** ANOM: GG AEF: 3H 7: Antonetti to Lasnet, August 24, 1929.

252 **"as soon as any civilized country":** *Proceedings of the International Labour Conference,* 12th sess. (Geneva: International Labor Organization, 1929), 1:44.

253 **"died off like flies":** Ibid., 1:63.

252 **the speaker had strayed beyond:** Ibid., 1:69.

253 **French Equatorial Africa could boast myriad:** ANOM 2H 3: *Enquête du BIT sur la main-d'oeuvre coloniale,* 1927.

253 **"the necessities of civilization":** *Proceedings of International Labour Conference,* 1:50.

254 **"in all its forms within":** International Labour Organization, *Forced Labour Convention, C29,* June 28, 1930. On British debates over forced labor and imperial legitimacy, see Kevin Grant, *A Civilised Savagery: Britain and the New Slaveries in Africa, 1884–1926* (New York: Routledge, 2005).

254 **"forced":** AILO: N206.1.106.1: Memos.

255 **improving lives through the drafting of regulations:** See J. P. Daughton, "ILO Expertise and Colonial Violence in the Interwar Years," in *Globalizing Social Rights: The International Labour Organization and Beyond,* ed. Sandrine Kott and Joëlle Droux (New York: Palgrave Macmillan, 2013), 85–97.

255 **"usefully consulted":** ANOM: GG AEF: 2H 3: Acting Governor-General J.-F. Reste to minister of colonies, June 7, 1927.

## CHAPTER 11: THE VICTORY AND THE FORGETTING

**258 "offensive":** Pierre Mille, "La 'Bataille' de Chemin-de-fer," *La Dépêche coloniale et maritime,* March 9, 1927.

**258 "the equatorial forest and the Mayombe":** SCB, *L'Oeuvre d'un siècle: 1846–1946* (Paris: Tolmer, 1952), 78, 80.

**259 Constant modifications of the route meant that:** "Les Voies de pénétration en A.E.F.," *L'Afrique équatoriale française,* special issue of *L'Orientation économique et financière* (September 13, 1930): 16.

**260 The broken landscape of the southern Middle Congo:** For discussion of the technical aspects of the construction, see the issue of *Le Génie civil* (July 14, 1934) on the Congo-Océan.

**261 In 1933 Mr. Girard, a lead Batignolles engineer:** "Courrier de l'Afrique équatoriale," *Les Annales coloniales,* April 1, 1933.

**263 One enthusiastic publication predicted:** "Les Voies de pénétration en A.E.F.," 16.

**263 never imported anywhere near 50,000 metric:** Raymond Susset, *La Vérité sur le Cameroun et l'Afrique équatoriale française* (Paris: Nouvelle revue critique, 1934). Susset's critiques were also reproduced in Jules Alcandre, "Le Désastre financier de l'Afrique équatoriale française," *Europe-colonies* (August 1934).

**264 an economic downturn since 1928:** Catherine Coquery-Vidrovitch, "L'Afrique coloniale française et la crise de 1930: crise structurelle et genèse du Rapport d'ensemble," *Revue française d'histoire d'outre-mer* 63, no. 232–33 (1976): 386–424.

**264 If the Congo-Océan had been meant to release:** On imports and export for 1935–36, see *Bulletin économique de l'Afrique equatoriale française* (Paris: A. Tournon, 1937).

**264 In her detailed history:** Anne Burnel, *La Société de construction des Batignolles de 1914 à 1939: histoire d'un déclin* (Geneva: Librairie Droz, 1995), 196–98.

**265 It was not a negligible amount of money:** Gilles Sautter, "Notes sur la construction du chemin de fer Congo-Océan (1921–1934)," *Cahiers d'études africaines* 7, no. 26 (1967): 288.

**265 Nearly forty years after its completion:** "Appraisal of a Railway Project, People's Republic of Congo," March 1, 1972, p. 7, http://documents1.worldbank.org/curated/en/244591468770719848/text/multi-page.txt (accessed July 2020).

**265 The use of concrete, prone to weakness and deterioration:** Ibid.

**266 Between 1990 and 2019, multiple accidents:** Congolese Embassy (France), "Le Chemin de fer Congo-Océan de 1990 à 2018, entre accidents, attentats, et guerre," *Actualités* (news service), November 26, 2018, Ambacongofr.org. These incidents were also covered in various media outlets.

**266 In 1984 Congolese president Denis Sassou-Nguesso:** Quoted in Sennen Andriamirado, *Le Défi du Congo-Océan; ou, l'épopée d'un chemin de fer* (Brazzaville: Agence transcongolaise des communications, 1984), viii.

**267 "free and friendly cheer":** Géorge E. Caillet, in *Étoile de l'A.E.F,* July 19, 1934, 7.

**267 a commemorative brochure:** *Inauguration du chemin de fer Congo-Océan. Pose de la prémiere pierre du Port de Pointe-Noire* (c. 1934), 3–4.

**267 "a revelation":** "Inauguration officielle du Congo-Océan," *Étoile de l'A.E.F,* July 12, 1934, 1.

**267 the focus of the festivities was on French:** *Étoile de l'A.E.F,* July 19, 1934, 2.

**268 "The penetration of the Continent":** *Inauguration du chemin de fer Congo-Océan,* 8–9.

268 **"the great works honoring the French":** Laval quoted in Burnel, *Société de construction des Batignolles*, 197.
268 **"separated from our *Patrie*":** *Inauguration du chemin de fer Congo-Océan*, 15.
269 **"remind us all of how much":** Ibid., 17.
269 **"mess—disorder—the impossibility":** Ibid., 19–20.
270 **"nuclei of white colonization":** Ibid., 24–27.
270 **"rough and murderous period":** Ibid., 27.
270 **The inauguration coincided with the publication:** République française, Afrique équatoriale française, *Le Chemin de fer Congo-Océan* (Paris: France Affiches, 1934).
271 **"slitting the throat of Africa":** Albert Londres, *Terre d'ébène* (1927; Paris: Collection Motifs, 2006), 240.
273 **René Maran did not stop criticizing:** Michel Fabre, "René Maran, the New Negro, and Negritude," *Phylon* 36, no. 3 (1975): 340–51.
274 **"an incomprehensible totalization":** Dr. Ferris quoted in Sautter, "Notes sur la construction," 268.
274 **some 16,000 men and women died:** Ibid., 269.
275 **in stories shared around the fire:** Francois Bita in Andriamirado, *Défi du Congo-Océan*, xiv.

## CHAPTER 12: THE VIOLENCE OF EMPIRE

277 **ideas about race:** Influential collections on how race and colonial power were constructed include Anne Stoler and Frederick Cooper, eds., *Tensions of Empire: Colonial Cultures in a Bourgeois World* (Berkeley: University of California Press, 1997); Julia Ann Clancy-Smith and Frances Gouda, eds., *Domesticating the Empire: Race, Gender, and Family Life in French and Dutch Colonialism* (Charlottesville: University of Virginia Press, 1998); and Sue Peabody and Tyler Edward Stovall, eds., *The Color of Liberty: Histories of Race in France* (Durham, N.C.: Duke University Press, 2003).
277 **the texture of racial distinctions:** See, for example, Jennifer Anne Boittin, Christina Firpo, and Emily Musil Church, "Hierarchies of Race and Gender in the French Colonial Empire, 1914–1946," *Historical Reflections* 37, no. 1 (2011): 60–90. On racial hierarchies in equatorial Africa, see Eric Jennings, *Free French Africa in World War II: The African Resistance* (Cambridge, U.K.: Cambridge University Press, 2015).
278 **"one doesn't treat a child":** ANOM: GG AEF: 3H 7: Antonetti to Lasnet, August 24, 1929.
278 **"this country to its primitive barbarism":** Antonetti quoted in J. Coupigny's preface to Michel R .O. Manot, *L'Aventure de l'or et du Congo-Océan* (Paris: Librairie Secretan, 1946), 9.
278 **workers as *nègres*:** On the history and contested meaning of the term *nègre*, see Brent Hayes Edwards, *The Practice of Diaspora: Literature, Translation, and the Rise of Black Internationalism* (Cambridge, Mass.: Harvard University Press, 2003), 26–35.
278 **black intellectuals in France had started to reclaim:** Jennifer Boittin, "Black in France: The Language and Politics of Race in the Late Third Republic," *French Politics, Culture and Society* 27, no. 2 (2009): 23–46.
278 **Africans found the term *nègre*:** Gaston Muraz, *Sous le grand soleil, chez les primitifs* (Coulommiers: Imprimerie Paul Brodard, 1923), xi–xii.

**278 "quarrelsome, invading":** Maurice Briault, *Les Sauvages d'Afrique* (Paris: Payot, 1943): 37.

**279 "These races die":** ANOM: FM AFFPOL 2856: M. R. Antonetti, *Discours prononcé à la séance d'ouverture du Conseil de gouvernement, session ordinaire de décembre 1925* (Brazzaville: Gouvernement général, 1926), 20.

**279 that most repugnant of customs, cannibalism:** Florence Bernault, *Colonial Transactions: Imaginaries, Bodies, and Histories in Gabon* (Durham, N.C.: Duke University Press, 2019), chap. 5; and Sophie Dulucq, "L'Imaginaire du cannibalisme: anthropophagie, alimentation et colonisation en France à la fin du XIX<sup>e</sup> siècle," *Se nourrir. Pratiques et stratégies alimentaires* (2013), https://hal.archives-ouvertes.fr/hal-00963880.

**279 "reserves of fattened children and slaves":** G. Renouard, *L'Ouest africain et les missions catholiques: Congo et Oubangui* (Paris: Oudin, 1904), 84, 98. The review is from *La Revue de géographie* 29, no. 15 (1905): 96.

**279 missionary journals referenced:** *Missions catholiques*, no. 3078 (January 18, 1929): 35.

**279 photograph allegedly of four cannibals:** Briault, *Les Sauvages d'Afrique*, plate VI.

**279 "would hunt whites":** M. R. Antonetti, *Discours prononcé à la séance d'ouverture du Conseil de gouvernement (Novembre 1928)* (Brazzaville: Gouvernement général, 1928), 29.

**280 to hang a sign warning guests:** Camille Drevet, *Les Annamites chex eux* (Paris: Société nouvelle d'éditions franco-slaves, 1928), 20.

**280 "I constantly saw Frenchmen":** Félicien Challaye, *Un livre noir du colonialisme: "Souvenirs sur la colonisation"* (Millau: Nuits Rouges, 2003), 34–35.

**280 "They have all the faults":** Robert Poulaine, *Étapes africaines: voyage autour du Congo* (Paris: Nouvelle revue critique, 1930), 45.

**280 "The contempt":** ALDH: F Delta rés 798/90: *Indigènes* of Lambaréné to administrative affairs inspector, March 5, 1916.

**281 "the *nègre* before the white man":** ALDH: F delta rés 798/90: *Indigènes* of Lambaréné to LDH president, December 20, 1916.

**281 "if Monsieur Gamon doesn't like":** ALDH: F Delta rés 798/179: Ernest Ogandago on behalf of "la jeunesse de Port Gentil," to governor-general, report, August 1, 1929.

**282 lack of "proof":** ALDH: F Delta rés 798/179: Raoul Mary to LDH, January 6, 1930.

**282 an allegedly liberal republican empire:** David Livingstone Smith, "Paradoxes of Dehumanization," *Social Theory and Practice* 42, no. 2 (2016): 416–43.

**282 Catholic missionaries, for example, complained:** See the correspondence and annual reports from Middle Congo at Archives de la Congrégation du Saint-Esprit: Congo: 3J3.2a6 and 3J3-.2a7.

**283 "disproportionate":** ANOM: AFFPOL 664: "Circulaire," March 14, 1935.

**283 the administration chose to ignore cases of abuse:** *L'Ouest africain français,* January 29, 1921, clipping in ALDH: F Delta rés 798/90.

**283 "negrophobia":** Ibid.

**284 "too rough":** ANOM: FM AFFPOL 2856: Labor Service director [Marchand], *Rapport sur le fonctionnement du Service de la main d'oeuvre du chemin de fer Congo-Océan (de janvier à avril 1925),* 7–8.

**285 The results of brutalization can include torture:** See, for example, John Dower, *War Without Mercy: Race and Power in the Pacific War* (New York: Pantheon, 1987); George Mosse, *Fallen Soldiers: Reshaping the Memory of the World Wars* (Oxford: Oxford

University Press, 1991); Christopher Browning, *Ordinary Men: Reserve Police Battalion 101 and the Final Solution in Poland* (New York: HarperPerennial, 1998); and Joanna Bourke, *An Intimate History of Killing: Face to Face Killing in Twentieth Century Warfare,* rev. ed. (New York: Basic Books, 2000).

285 **they broke psychologically:** Without discussing "brutalization" per se, Patrick Brantlinger nonetheless explores the discourse of Europeans living in such trying colonial conditions in *Dark Vanishings: Discourses on the Extinction of Primitive Races 1800–1930* (Ithaca, N.Y.: Cornell University Press, 2003).

285 **tribulation of remote travel:** Johannes Fabian, *Out of Our Minds: Reason and Madness in the Exploration of Central Africa* (Berkeley: University of California Press, 2000): 145.

285 **white overseers sometimes encountered:** See, for example, Manot, *L'Aventure de l'or,* 62–65.

285 **The European driven mad by the tribulations:** Thomas de Quincey, *Confessions of an English Opium Eater* (London: Taylor and Hessey, 1823); and Doris Lessing, *The Grass Is Singing* (London: Michael Joseph, 1950).

285 **He represents the promise and threat of empire:** Joseph Conrad, *Heart of Darkness* (1899; New York: W. W. Norton, 1988), 28.

285 **"a thoroughgoing racist":** Chinua Achebe, "An Image of Africa: Racism in Conrad's *Heart of Darkness,*" *Massachusetts Review* 57, no. 1 (2016): 21. For a discussion of Achebe's reading of Conrad, see Patrick Brantlinger, *Rule of Darkness: British Literature and Imperialism, 1830–1914* (Ithaca, N.Y.: Cornell University Press, 1990).

285 **"the twentieth century's most famous":** Adam Hochschild, *King Leopold's Ghost: A Story of Greed, Terror, and Heroism in Colonial Africa* (New York: Mariner Books, 1999), 144.

286 **Conrad's "real life" inspirations:** Hannah Arendt linked Kurtz to the German Carl Peters; see Arendt, *The Origins of Totalitarianism* (San Diego, Calif.: Harcourt, 1968), 189. Adam Hochschild posits a number of real figures, including Léon Rom of the Force publique; see Hochschild, *King Leopold's Ghost,* 144–45.

286 **African colonies were a place:** Arendt's section on imperialism in *The Origins of Totalitarianism* has been often discussed by intellectual historians, political theorists, and others. See, for example, Richard King and Dan Stone, eds., *Hannah Arendt and the Uses of History: Imperialism, Nation, Race, and Genocide* (New York: Berghahn Books, 2007); and Benjamin Claude Brower, "Genealogies of Modern Violence: Arendt and Imperialism in Africa, 1830–1914," in *The Cambridge World History of Violence,* vol. 4, *1800 to the Present,* ed. Louise Edwards, Nigel Penn, and Jay Winter (Cambridge, U.K.: Cambridge University Press, 2020), 246–62.

286 **"felt not only the closeness of men":** Arendt, *Origins of Totalitarianism,* 184.

287 **allowed a white man to be "a chief":** William Cohen, "The Lure of Empire: Why Frenchmen Entered the Colonial Service," *Journal of Contemporary History* 4, no. 1 (1969): 103–16.

287 **"I am chief because I have":** Delavignette quoted ibid., 111.

287 **"strains the nerves":** Georges Hardy, *Nos Grands Problèmes coloniaux* (Paris: Albert Colin, 1929), 123–24.

288 **"anguish of nights deep":** Ibid., 143–44.

288 **"a vigorous morality"**: Ibid., 145–46.
288 **"of a sure morality, well paid"**: ANOM: GG AEF: 3H 8: Marchand, "Observations," c. early 1925.
289 **"but they have only hatred"**: Sauvage, *Secrets de l'Afrique noire*, 313.
289 **disdain for Spaniards, Italians, and Portuguese:** See, for example, Ralph Schor, *L'Opinion française et les étrangers, 1919–1939* (Paris: Publications de la Sorbonne, 1985).
289 **faced daunting challenges:** Poulaine, *Étapes africaines*, 70.
290 **"like the natives"**: ANOM: FM AFFPOL 2856: *Note pour l'instruction générale des travaux publics*, May 22, 1933.
291 **"unbalanced and even criminal agents"**: ANOM: GG AEF: 5D 60: Antonetti to minister of colonies, January 23, 1926.
292 **"these most corrupt agents"**: Ibid.
294 **administration's culpability for delays:** ANMT: 89 AQ 873: SCB Correspondence, April 7, 1926–February 29, 1928, *Tableau donnant mensuellement les effectifs journaliers demandés, ceux obtenus, et leur différence*.
294 **"Believe me, Monsieur"**: Rouberol to Antonetti, August 19, 1926, ibid.
295 **"exceeded the possibilities of the colony"**: Minister of colonies, in *Journal officiel de la République française. Débats parlementaires. Chambre des députés*, November 23, 1927, 3185.
295 **"turned this poor colony upside down"**: M. Dubosc, elected delegate to Conseil supérieur des colonies, quoted in "Pour Vous, M. Millies-Lacroix," *Capital et travail*, May 20, 1926.
295 **the Kongo-Wara rebellion:** Thomas O'Toole, "The 1928–1931 Gbaya Insurrection in Ubangi-Shari: Messianic Movement or Village Self-Defense?" *Canadian Journal of African Studies* 18, no. 2 (1984): 329–44.
295 **repeatedly downplayed the uprising:** ANOM: GG AEF: 5D 47: Antonetti to minister of colonies, June 8, 1929.
295 **"Since Makoua, the villages"**: H——, "Tournée du 21 novembre au 21 décembre 1927," included as an appendix in Poulaine, *Étapes africaines*, 229–31.
296 **rebellious villagers killed chiefs and soldiers:** "Que se passe-t-il en Oubangui-Chari," *Les Annales coloniales*, January 10, 1929.
296 **"never been established"**: ANOM: GG AEF: 5D 47: Marchessou, report, October 25, 1929, 4, 61–62.
298 **"Wonder has been expressed"**: Albert Memmi, *The Colonizer and the Colonized*, trans. Howard Greenfield (New York: Orion Press, 1965), 21.
298 **"which depend themselves on the construction"**: ANOM: GG AEF: 5D 60: Antonetti to LDH president, February 20, 1927.
298 **"I first have to remind you"**: ANOM: GG AEF: 5D 60: Marchand to Lower Ubangi district head, February 24, 1925.
299 **responsibility would fall on the colony:** A number of social psychologists have come to similar conclusions when studying the functioning of bureaucracy. See, e.g., Nestar J. C. Russell and Robert J. Gregory, "Spinning an Organizational 'Web of Obligation'? Moral Choice in Stanley Milgram's 'Obedience' Experiments," *American Review of Public Administration* 41, no. 5 (2011): 495–518.
299 **regional officials enjoyed total power:** René Maran, "La Grande Souffrance des nègres de l'Afrique équatoriale française," *Le Peuple*, August 20, 1933.

299 **lack of oversight:** On the shortcomings of the colonial administration in Africa, see William B. Cohen's wonderful study, *Rulers of Empire: The French Colonial Service in Africa* (Stanford, Calif.: Hoover Institute Press, 1971).

299 **being "tough":** These terms were used in correspondence read aloud in the National Assembly. *Journal officiel de la République française. Débats parlementaires. Chambre des députés,* November 24, 1927, 3177.

299 **"native mentality":** ANOM: GG AEF: 5D 47: Antonetti to minister of colonies, June 8, 1929.

301 **"a significant reduction in the performance of those who come to work":** AMT: 89 AQ 873: Lebert to Antonetti, April 22, 1926, quoted in Anne Burnel, *La Société de construction des Batignolles de 1914 à 1939: histoire d'un déclin* (Geneva: Librairie Droz, 1995).

302 **two slim columns:** See, for example, AMT: 89 AQ 917: XXXIII: Rapports hebdomadaires: Brigades d'études.

302 **"no means of coercion":** AMT: SCB: 89 AQ 873: Lebert to Antonetti, September 10, 1925.

303 **"a black man wilts":** Albert Londres, *Terre d'ébène* (1927; Paris: Collection Motifs, 2006), 262.

303 **"weaken, fall prey to nostalgia":** ANOM: GG AEF: 3H 8: Boyé, *Rapport médical annuel (octobre 1924–décembre 1925).*

303 **"incorrigible uncleanliness":** ANOM: GG AEF: 3H 8: [Lefrou,] *Chemin de fer Brazzaville-Océan, division côtière, ambulance de Mavouadi, compte rendu mensuel, année 1925, mois de juin.*

304 **"state of physical and moral inferiority":** ANOM: GG AEF: 3H 8: Boyé, *Rapport médical annuel (octobre 1924–décembre 1925).*

304 **violence and indifference stemmed:** See Hannah Arendt, *On Violence* (New York: Harvest Books, 1970).

305 **alliances, agreements, taxation, and terror:** On politics in this region, see Jan Vansina, *Paths in the Rainforests: Towards a History of Political Tradition in Equatorial Africa* (Madison: University of Wisconsin Press, 1990); and Vansina, "The Peoples of the Forest," in *History of Central Africa,* ed. David Birmingham and Phyllis M. Martin (New York: Longman, 1983), 1:75–117.

# ILLUSTRATION CREDITS

86　Congo Français, "Passage de M. l'Administrateur E.," postcard, c. 1905, Jean Audema, Eliot Elisofon Photographic Archives, National Museum of African Art, Smithsonian Institution.

95　Workers from Chad or Ubangi-Shari, with a French administrator, c. 1926: Courtesy of the Department of Special Collections, Stanford University Libraries.

116　Route and corresponding altitude of the Congo-Océan railroad: Map based on *Le Génie civil*, July 14, 1934.

118　"Dans le Mayumbe – Congo français," c. 1910: postcard. Jean Audema.

131　"Type de femme à M'Boukou," c. 1925. Reproduced with the permission of the Archives nationales du monde du travail, Roubaix (89 AQ 1000: Photographies).

173　"Mavouadi Kil 61: Equipe des travailleurs de la Brigade d'Etudes," c. 1925. Reproduced with the permission of the Archives nationales du monde du travail, Roubaix (89 AQ 1000: Photographies).

202　"Mavouadi Kil 61. Equipe de travailleurs malades à leur arrives sur les travaux," June 1925. Reproduced with the permission of the Archives nationales du monde du travail, Roubaix (89 AQ 1000: Photographies).

260　"Tunnel de 109—côte Brazza," September 1932. Reproduced with the permission of the Archives nationales du monde du travail, Roubaix (89 AQ 1000: Photographies).

262　"Viaduc de 108," September 1932. Reproduced with the permission of the Archives nationales du monde du travail, Roubaix (89 AQ 1000: Photographies).

272　"Le Mayombe après les travaux," from *Le Chemin de fer Congo-Océan* (Paris: France Affiches, 1934). Courtesy Bibliothèque Nationale de France.

291　"Entrée du souterrain du Bamba, Km. 141," c. 1933. Reproduced with the permission of the Archives nationales du monde du travail, Roubaix (89 AQ 1000: Photographies).

# INDEX

Page numbers in *italics* refer to illustrations. Page numbers beginning
with 313 refer to notes.